LIFE AS IT IS

LIFE AS IT IS

BIOLOGY FOR THE PUBLIC SPHERE

WILLIAM F. LOOMIS

UNIVERSITY OF CALIFORNIA PRESS
BERKELEY   LOS ANGELES   LONDON

University of California Press, one of the most distinguished
university presses in the United States, enriches lives around
the world by advancing scholarship in the humanities, social
sciences, and natural sciences. Its activities are supported by
the UC Press Foundation and by philanthropic contributions
from individuals and institutions. For more information, visit
www.ucpress.edu.

University of California Press
Berkeley and Los Angeles, California

University of California Press, Ltd.
London, England

Library of Congress Cataloging-in-Publication Data

Loomis, William F.
    Life as it is : biology for the public sphere / William F. Loomis.
        p.   cm.
    Includes bibliographical references and index.
    ISBN: 978-0-520-26001-6 (pbk. : alk. paper)
    1. Biology—Social aspects—Popular works.   2. Life
sciences—Popular works.   I. Title.
QH333.L66   2008
612—dc22   2007027034

Manufactured in the United States of America

17   16   15   14   13   12   11   10   09
10   9   8   7   6   5   4   3   2   1

The paper used in this publication meets the minimum
requirements of ANSI/NISO Z39.48–1992 (R 1997) (*Permanence
of Paper*).

TO MY WIFE MARGA
AND
MY DAUGHTER CAROLINA

CONTENTS

ILLUSTRATIONS

PREFACE

PEOPLE HAVE DISCUSSED THE NATURE OF LIFE SINCE TIME IMMEMO-
rial. What makes it special, how is it passed on to new generations, and what happens when it ends? Recently there has been
a dramatic increase in our understanding of the cellular basis
of life, of the role of DNA sequences in determining traits, and
of the biological basis for consciousness, and we are now better
able to manipulate genomes. When I started this book, it
seemed to me that the political controversies concerning abortion, euthanasia, and the establishment of embryonic stem cells
(ES cells) were out of touch with biological reality. They were
distracting from the far more serious societal problems of
human evolution and population increase. Since these are emotional matters, it was essential to put them in human terms.
Religious and governmental stances had to be integrated into
the presentation while avoiding dogmatic positions. Although
the moral questions are clear, their answers depend on context
and personal values. I have attempted to present the problems
in a way that permits each individual to reach his or her own
conclusions.

From a cell biologist's point of view, life is cheap. Cells grow rapidly when given the proper nutrients and are really only of interest to scientists. Human cells growing in the lab are not significantly different from mouse cells and do not get any special respect. Likewise, eggs of any species start to divide soon after fertilization and form blastocysts with hundreds of cells. Only when the fertilized eggs come from humans is the question of dignity raised. The cells' potential to become an individual is there, but when is it actualized? Many societies seem to be quite confused about the answer. In no society is a pregnant woman counted as two in a census of the population. The fetus is considered a citizen only after birth. Yet in many societies there are restrictions on late-term abortions even when the mother's health is at risk. And many societies do not condone the establishment of ES cells, because an embryo is sacrificed. To say that cells are cheap can inflame passions, because it resonates with the old stories that barbarian potentates considered life cheap. Genghis Khan may have felt that killing hundreds of thousands was a small price to pay for dominion over the world, but we think we are more civilized now. However, we should not confuse our compassion for our fellow human beings with feelings for simple cells. Compassion should be focused on nurtured human life, the most precious thing in the world.

To make the case, I felt that it was necessary to point out the shared aspects of all living things and then try to establish what is special about human life. This led me to considerations of individuality, consciousness, and social behavior unique to humankind. I used a graded scale in discussing core consciousness, awareness, attention, and extended consciousness. The role of language in generating the internal narrative used in decision making and planning is uniquely human. The questions raised concerning the interactions of individuals, tribes, and civilizations do not have clear-cut biological answers; we have only generalities and plausible explanations with which to respond to them. Nevertheless, these questions form the basis for societal decisions that can affect us all.

Evolution is woven throughout the presentation because, as the eminent geneticist Theodosius Dobzhansky famously said, "in biology nothing makes sense except in the light of evolution." The working of chance and necessity on inherited characteristics is now well understood. It may be humbling to realize that we are the way we are as a consequence of random changes in the sequence of DNA that every so often resulted in selective advantages for one lifestyle or another, but this is what the DNA says. To push home the point that evolution can account for all the living things around us, I even took up the most poorly understood step in the history of life on earth—the origin of life. We may not fully understand how a self-sustaining system could evolve from the simple molecules present on the young planet, but we understand enough to know that there is no reason to think it implausible. There is no evidence that supernatural processes were required, just a lot of trial and error over millions of years. Once life got going, nothing could stop it, and all sorts of wondrous organisms arose. Most went extinct after a while, but many others continued to evolve, until there were primates, great apes, and, finally, humans. The story gets more complete every year as fossils are found all over the world and comparative genomics draws a thread through them all.

We humans are a special species because we have learned to control the world around us to a much greater extent than any other species. We build dams and irrigate deserts, and grow more food than we need. We exploit natural resources such as coal and iron to build cities of steel and ships to carry the produce around. Meanwhile, we have multiplied and filled the earth with more people than it can support and are now subjecting our resources to deficit spending. The wealthy nations of North America and Europe use more than their fair share of the resources and produce the bulk of the greenhouse gases, such as carbon dioxide, that are leading to global warming. The Kyoto Protocol of 2005 requires participating developed countries to reduce their emission of greenhouse gases to below 1990 levels by 2012. However, the largest producer of greenhouse gases, the United States, has not ratified the treaty, partly because China, the second-greatest emitter of greenhouse gases, is exempt from the restrictions since it is considered a developing country.

Economic and political competition make it unlikely that the goals will be reached before temperatures increase several degrees, disrupting weather patterns and causing the sea level to rise dramatically. Millions will be displaced, and the political stability of the world threatened. Such rapid change is likely to disrupt commerce and lead to famines. It is fairly clear that the global population will decrease drastically in the near future; the only question is how the decrease will come about. I argue as strongly as possible that the most acceptable way to bring about a significant drop in world population is to have fewer children over the next several generations. Once the global population returns to the level of a hundred years ago, the birthrate could be allowed to gradually increase to establish a new, steady state. Only a full commitment to such a course in the near future has any hope of succeeding. The alternatives are all rather ghastly.

But all is not doom and gloom. Humans are resilient, resourceful, and surprisingly responsible. As the fruits of biological and sociological research become public knowledge, people can be counted on to confront the problems and plan for the future. Our ability to understand and alleviate devastating diseases will increase enormously if we are able to follow the courses of specified diseases in genetically defined cells growing in the controlled environment of a laboratory. When we learn to direct the differentiation of healthy ES cells to specific cell types, we can use them to replace degenerating tissues in otherwise untreatable cases. The new biology may also be harnessed to reduce the birthrate, so that we can look forward to a healthier, more sustainable population in generations to come. Treating the sick while also working to reduce global population is not a contradiction: we can help nurture life while avoiding stressing the planetary resources to the breaking point.

The new perspectives that come from a better understanding of genetics and genomics do not diminish individual dignity but allow each of us to recognize our possibilities and our limitations. Many of our nonconscious reactions may be determined by mental modules that have been selected for survival through our long evolution from the early mammals. However, we accept responsibility for our actions and decisions under the assumption that they result from free will. We carry on an internal dialogue in our minds using that uniquely human ability, language. Recent advances in communication technology allow us to interact continuously with people all over the world by telephone, Internet, and overnight flight. It is becoming increasingly clear that we are mutually dependent and share the same small planet.

Many of the subjects covered in this book are controversial, but we can benefit from informed discussions about them from a biological perspective. I have had a lifelong interest in questions concerning directed evolution, selfishness, morality, population control, and other topics that affect us all. I have followed the progress made in diverse fields in clarifying thinking about the underlying biological processes and the sociological consequences of intervention. Examining the questions from a biological perspective has helped me find rational answers. New results continue to lead to new insights that will undoubtedly affect future considerations. However, the world is confronted today by tough questions that will affect generations to come. A consensus on these questions must be reached soon, and decisions must be made. Science is moving too fast for us to put off the problems for later. It would be best to approach them openly on the basis of all the available information.

I have spent most of my life immersed in the molecular details of the development and evolution of a simple soil amoeba, and I have benefited from continuous discussions with brilliant specialists in other fields. There are too many to list, but all are held dearly in my memory. In recent years I have benefited from many talks with my friend and colleague Philip Kitcher, who moved a few years ago to the Department of Philosophy at Columbia University. He showed me the complexities of the philosophy of science. My wife, Margarita Behrens, has patiently explained the subtleties of neurobiology and cheered me on as I have tried to apply them to psychological situations. She also came up with the title that stuck to this book. Henri Korn, an academician at the Institute Pasteur, convinced me to draw a thicker thread through the various chapters. I received good advice from many friends and family members who read various drafts, and I thank them all. I am grateful to Chuck Crumly, my editor at University of California Press, for guiding the book through to publication.

I have tried to make each chapter stand on its own, but the book is better if read from the beginning. Throughout the text I have indicated recent books and papers that give primary data and deeper discussions. Further information and different points of view can be found in essays referenced in those sources. I wrote this book with the nonspecialist layperson in mind, but I hope that the specialists do not find any glaring errors.

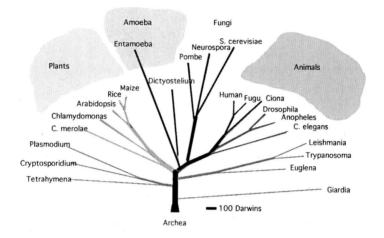

ONE

THE VALUE OF LIFE

FOR THE LAST HUNDRED YEARS, WE HAVE HAD FRONT ROW SEATS AT one of the greatest shows on earth—the birth and flowering of modern biological sciences. The insights and techniques for understanding the roles of specific macromolecules in the functions of cells and organisms have transformed biology from a descriptive to a predictive science. Over the years, universal rules have been proposed, considered, rejected, modified, and accepted. The baroque activities of cells during the development of embryos have been laid bare by molecular color coding of each cell type as it arises. Embryogenesis is no longer a mysterious process with long lists of names for the different structures we see in fish, birds, and mammals. It is a play of genes and proteins, and in it the potential that lies in every fertilized egg unfolds.

FIGURE 1.0 Radiation of the eukaryotes. This tree is rooted on a set of archaebacteria. Single-celled protists diverged before the split between plants and animals. The positions of the nodes and the branch lengths are quantitatively determined by genomic comparisons.

When we can see each act of the play, we learn about ourselves, for each of us was once a fertilized egg. The principal actors are the ancient genes that evolved long ago in other roles, and which have taken on new parts by subtle changes in their makeup. The realization that all life is connected at the basic level changes how we think about life. We can learn about ourselves by studying simpler life-forms, testing treatments in model organisms, and treating cells for what they are, rather than what they might become. We can consider how the brain develops emotions, feelings, thoughts, and memories, and how it generates a sense of self and a feeling of responsibility. The science of mind can be used to tackle the problems that come from having such an overdeveloped organ as the brain.

We are the only beings on the planet who carry on a continuous internal conversation with ourselves, the only ones with a movie-in-the-brain that plays images of people and things from the past, the present, and the future. It makes us a curious species with high self-regard. And because we think we're special, we often lose sight of our basic nature. In writing this book, I have tried to use modern biological insights to think sensibly about a whole lot of things—birth, death, cloning, abortion, euthanasia, evolution, individuality, consciousness, and morality. Only when we see life on earth as a grand and precarious creation that we are a part of, not apart from, will we be able to pass on our resources and knowledge to future generations.

Modern biology has also raised thorny questions as never before. Like the rebel god Prometheus, molecular biology has brought us the power to light up the darkness or burn down the house. Knowing how genes work opens up avenues for a healthier life, or it can shake the foundations of our society. The ability to clone both genes and organisms has begun a new era in biology. There are no rebel gods in this story, although there are heroes and a few charlatans. Mostly this book is an attempt to offer a reasoned discussion of the difficult problems facing every society given the new knowledge. Forethought (which is the translation of the name Prometheus) can be used to see the road ahead and to attempt to avoid the worst of the potholes. However, we should never lose sight of the preciousness and excitement of life. It is the greatest adventure we will ever know.

THE NATURE OF LIFE

Life is everywhere—in the air, on the soil, in the oceans. Birds fly, mammals run, fish swim, plants grow tall. But by almost any way of counting or weighing, most living matter on this planet is made up of organisms that grow and divide as single cells too small to see with the naked eye. They are the bacteria and algae that have flourished in the seas for billions of years. Over the first two billion years, they slowly changed the very chemistry of the earth's surface, so that the atmosphere became rich in oxygen. Once this happened, most of these early organisms had to retreat to deep, dark regions that were still anaerobic, but a few were able to survive the toxicity of oxygen and live near the surface. They evolved the ability to use oxygen's reactive properties to burn nutrients more efficiently and generate energy at a high rate. These aerobic organisms gave rise to multicellular organisms that could use their size to graze on smaller cells. Organisms that you could see with the naked eye appeared about a billion years ago and gave rise to the animals and plants that we have around us now. Only in the last 10 million years did the great apes come out of the primate line, and one of them evolved into us, *Homo sapiens*. The children of primates are born almost defenseless and need to be nurtured for a long period before they are able to fend for themselves. Nurtured life is one of the most precious things in the world. But, given all the bacteria and algae that divide to give rise to new cells every day, life itself is cheap.

It is only we humans who ask, "What is life?" We ask when life begins—Is it when an egg is fertilized by a sperm? When the egg divides? When it gives rise to an embryo that can live on its own? Or only when it is born? We ask, when is life over—Is it when the heart stops? When all brain activity has ceased? Or when the probability of recovery is negligible? Or is it when the body grows cold and rigor mortis sets in? These are important questions, but they are asked as if only mammals mattered. Perhaps it would be useful to first think of life from the broader point of view of all organisms and then return to the especially human aspects of life.

"What is the characteristic feature of life? When is a piece of matter said to be alive?" wrote Erwin Schroedinger in his book *What Is Life?* His answer was, "When it goes on 'doing something', moving, exchanging material with its environment, and so forth, and that for a much longer period than we would expect an inanimate piece of matter to 'keep going' under similar circumstances" (Schroedinger 1944). Living things might continue to move after the point when a pebble would have stopped. They might continue to grow after the point when a gel would have swollen to its full extent. They might continue to divide in a regular fashion after the point when a sand castle would have fallen down. Dead things stop after a while, but living things keep on going. In fact, they have been growing, dividing, and changing since life first arose on this planet.

All living forms consist of single cells or groups of cells that take up matter or energy from the environment, grow, and divide to increase their number. Most of them can move in one way or another to find a better source of food or energy, or to fit better into the structure of a tissue. Cells are usually too small to see with the unaided eye, but a few become enormous. An ostrich egg is a single cell, but after fertilization it quickly divides up into millions of cells, so that each cell in an ostrich chick is about ten micrometers in diameter and could fit on the point of a needle. Bacteria are even smaller, some as small as one micrometer in diameter, which is about the theoretical limit for a cell. You need a good microscope to see bacteria.

All cells are surrounded by a membrane with a central layer of fatty acids that limits the flow of water and small molecules back and forth between the cell and the environment. These membranes give each cell a degree of individuality and privacy while not precluding cooperation. Chemical reactions go on inside the cellular membrane and convert substances taken up from the environment into needed molecules. The reactions that together form the central metabolic pathways interconnecting most of the common subunits are found in all cells. These reactions are catalyzed by a set of proteins that fold in a manner that permits them to distinguish among the small molecules and bind only to some and not others. Proteins are long chains of twenty to two hundred amino acids in a row, where any of twenty different amino acids can be found at any position. However, each protein has a unique sequence that determines the shape it will take up. If all possible sequences of one hundred amino acids are considered, there could be 20^{100} different sequences. But only a tiny fraction of the total number of possible sequences is found in any living being. Proteins able to catalyze the same chemical reaction in cells of different types often have very similar amino acid sequences.

For instance, a bacterial protein that catalyzes the interconversion of two small molecules during the fermentation of glucose has almost the same sequence as the protein that catalyzes the same reaction in humans. Either only a few of the 20^{100} possible sequences fold in the right way to catalyze this reaction, or bacteria and humans inherited the information to make this enzyme from a common ancestor and have not changed it much ever since. The first possibility can be ruled out, because we know of proteins in archaebacteria with sequences almost completely different from those in bacteria that catalyze this reaction. They take up the proper shape to distinguish the three-carbon compound from other very similar small molecules. The archaebacterial sequence has one short run of 5 amino acids that are in the same order as a run of amino acids in the bacterial enzyme, but other similarities in the sequence are few and far between. Although the archaebacterial and bacterial sequences may be distantly related, it is clear that proteins with little similarity in sequence can catalyze the same reactions. So why are almost half the amino acids in this 248-amino acid chain identical in the enzyme in humans and the enzymes in several bacterial species? A shared ancestral sequence that has been inherited for billions of generations is the likely answer.

A comparison of the sequence of this protein—which is called triose phosphate isomerase—found in rabbits and the same protein in humans shows the two proteins are 98 percent identical. Out of the 248 amino acids in a row, only 2 in rabbits, 4 in dogs, and 1 in chimpanzees differ from the human sequence. In fact, there is no question that all mammals inherited the ability to string amino acids together in the same sequence in this enzyme. Since 159 of the amino acids in triose phosphate isomerase are in the exact same order in some plants, and over 100 amino acids are identical in some bacteria, it appears that this is an ancient sequence inherited from a cell that long ago gave rise to not only bacteria but also plants and animals. This is not the only protein inherited from a common universal ancestor: thousands of proteins appear to be shared between plants and animals, and many of them clearly had a bacterial origin. The sequences vary to a certain degree in the different organisms, but they are all very similar, and the similarity could not have arisen twice by luck alone. The chance that they are unrelated is less than 1 in 10^{50}, a number so small that it can be safely ignored.

The evidence that all life on this planet is related goes on and on (Loomis 1988). Once life originated, it thrived, giving rise to unimaginably huge numbers of cells. Some of the progeny slowly changed and gradually produced all the diversity of life we see around us. There are several arguments that the first cells on earth were similar to the bacteria that live today. Rocks that are more than 3 billion years old show signs that they were made by deposits collected around colonies of bacteria. And there are fossil traces in other rocks of the same age that resemble strings of bacteria. Fossils of the kinds of cells that make up plants and animals first appear in rocks that are a billion years old. These cells, called eukaryotes, are characterized by being larger and carrying their chromosomes in a nuclear envelope. It is not surprising that bacteria preceded eukaryotes, since simpler cells with a single membrane on the outside would be expected to evolve before cells with a nucleus. It is likely that one of these simpler cells, which had had the world to themselves for so long, gave rise to a rare variant with internal compartments, which then shared the planet with them.

TIME AND DESCENT

Until recently, history was told in epic tales passed on within tribes. Elders would recite the stories they had heard as children around the campfires, and their children and grandchildren would listen and pass them on years later. The stories changed as old memories faded and recent events were more vividly recalled. By the time these stories were written down, starting about three thousand years ago, the origins of the tribe were clothed in poetry and given supernatural meaning. Some of the myths and stories seemed to stretch back to the beginning of time. Tribes that traced their lineage back for five hundred or a thousand years could conceive of a distant past when all things were new. A few thousand years seemed such an immense period of time that it could include the creation of the fish of the seas, birds of the air, and beasts of the land. Beautiful creation stories were written and recited.

For a long time the tales filled the need to account for the known world and all that was in it. But travelers came with seashells found in the rocks at the tops of mountains. How could they ever have gotten there? Fossils of organisms with forms never before seen started to pile up that required fanciful additions to the creation stories. It slowly dawned on people that the earth must be much older than a few thousand years. In the nineteenth century, geologists such as Charles Lyell realized that some rocks were very old. His friend Charles Darwin took a copy of his book *Principles of Geology* on the voyage of the *Beagle* in 1831. In this book Lyell argues that slow changes in the earth had raised seafloors into mountains and then eroded them away over millions of years. In 1867 Lyell estimated the start of the Ordovician period at 240 million years ago based on the fossil record. We now know it began about 500 million years ago, but it was not a bad guess.

In 1907, Bertram Boltwood came up with a whole new way of dating rocks. He realized that radioactive uranium spontaneously decayed through a series of elements until it turned into lead. The decay was slow and continuous and provided a sort of clock. The half-life of uranium^{238} could be measured, and was found to be 4.5 billion years. So the older the rocks, the more the uranium^{238} would have turned into lead^{206}. By measuring the content of lead in uranium ore that was available, Boltwood estimated the age of the earth to be at least 2 billion years. No attempts were made to find older ore at that time. When isotope chronology techniques were applied to a dozen or so meteorites, they were all found to be 4.5 billion years old. These rocks had been in cold, deep space, where nothing happened that could affect the decay of atoms or the retention of the daughter atoms. Their atoms had been decaying since the formation of the solar system. Various independent radiometric dating procedures, including decay of rubidium[87] to strontium[87] (half-life 49 billion years) and ratios of lead isotopes, have been applied to meteorites and ancient earth rocks and found to give almost identical ages.

The oldest rocks found on the surface of the earth have been dated to 4 billion years. They are found in the Acasta Gneiss rock outcrop in western Canada near Great Slave Lake. Very few out-croppings as old as these have been found elsewhere. However, there are rocks near Warrawoona in western Australia that date from 3.5 billion years ago, and some of these contain what appear to be fossil bacteria (Knoll 2003). In the last century, as we have come to realize that life on earth is billions, not thousands, of years old, it has become clear that we must learn to think across vast expanses of time that dwarf our concepts of history. Just as we have to think small when we look into a microscope to see the tiny bacteria that crowd every puddle, we must stretch our thoughts to encompass billions of years during which evolution slowly shaped the forms of life. From the beginning, life gave rise to life. Small changes led to cells better able to cope and multiply. Evolution continued at its slow pace for billions of years. It may be humbling to think that we are at the tag end of a very long story. But what a story it is. A single line goes on for four billion years!

Heredity insures that information available in one generation is passed on to the next, and that a variant trait reappears in subsequent copies so that it can be perpetuated in the lineage.

Natural selection determines whether or not the trait spreads through the population in generations to come. This is Darwin's descent with modifications.

The information necessary to string together a specific sequence of amino acids in a protein is encoded in the sequence of another class of long polymers, the nucleic acids DNA and RNA. These molecules are long chains made from only four different subunits. The nucleosides adenosine (A), thymine (T) (uridine in RNA), guanosine (G), and cytosine (C) are bound to each other by phosphate groups linking the sugar groups that are common to them all. The information is in the specific sequence of As, Ts, Gs, and Cs. The sequence is read out in groups of three, so that each of the twenty amino acids used in proteins can be specified. What is surprising is that the readout of the code is exactly the same in archaebacteria, bacteria, and all eukaryotes. The triplet ATG encodes one of the amino acids, methionine, in the bacterium *Escherichia coli*, the Archaea *Thermoplasma volcanium*, and humans. Moreover, the first amino acid in every protein found in living cells is methionine, and the first triplet, or codon, is always ATG. The other nineteen amino acids are encoded by other combinations of three nucleosides. The mechanism for determining where to start translating a nucleic acid begins with an ATG and then translates the rest of the sequence in groups of three bases. This holds for every bacterium that has been studied, as well as every plant and animal. Because the readout is the same in all organisms, human genes can be expressed in bacteria, and bacterial genes work fine in human cells.

The universality of the code, as well as details of the shared translation process by themselves, provides convincing evidence that all life is descended from a single cell that arose billions of years ago. The code itself appears to be a historical accident and could have been very different, but once it was in place in a successful cell, it could not change, since it is used in making every protein. Any change in a codon would result in a change in the amino acid sequence wherever the amino acid was specified. No cell could survive with changes in all its proteins. So we have all stayed with the same set of codons, although the sequence of amino acids differs to some extent in our proteins. This commonality is so striking that there is no question that all life on earth is related at this fundamental level. The nature of life can be considered for archaebacteria that live at volcanic vents under the sea just as much as for birds in the air.

Many plants and bacteria don't move on their own, and so we cannot look to movement alone as a sign of life. But they all grow and divide and have been doing so for billions of years. A microbiologist does not question whether a bacterium is alive or dead. Either it can give rise to progeny or it can't. When a population of living bacteria is diluted and spread on a suitable food source, each cell will give rise to millions that pile up into a colony that can be easily seen on a Petri dish. If the population has been harmed, say by irradiation with strong ultraviolet light or with X-rays, many cells will be unable to give rise to viable progeny, and the number of colonies will be far fewer than the number of cells plated. If the irradiation is carried on for long, there may be no colonies at all because all the cells will be killed. Immediately after lethal irradiation, many of the cells will continue to metabolize nutrients, generate the chemical energy (ATP) that powers many reactions, and even make new proteins. But their nucleic acids will have been irreversibly harmed, and any progeny cells generated by a last attempt at division will not grow. They will not be able to "keep going" any longer than an inanimate piece of matter under similar circumstances.

When a bird falls dead out of the air, there is little question that it is no longer alive. However, it may stay warm for some time since its cells are still able to metabolize and generate ATP. Some muscles may twitch for a while, but soon the body will grow cold and rigor mortis will set in. Were the cells still alive when the bird first fell? If one applied the microbiologist's criterion that they could give rise to viable progeny if spread on a suitable food source in a Petri dish, they have to be considered alive. But they are bird cells and not a bird. The bird is dead.

RANDOM MUTATIONS

The sequence of nucleosides in nucleic acids is copied before being passed on to progeny. DNA is a double-stranded helix with adenosine always found paired with thymine, and guanine found across from cytosine. Each strand has the same information, and one is the complement of the other. When they are both replicated, the information in both strands is exactly reproduced. Since only one strand directs the sequence of amino acids in proteins, there is no confusion about the code. However, if the sequence of nucleic acids in DNA were always inherited perfectly, only those proteins encoded in the first successful cell would be found in living things. Life would be restricted to bacteria-like organisms with no chance of change. However, on rare occasions, errors are made in the replication of nucleic acids that result in random changes.

The DNA replication machinery has a built-in proofreading function that immediately corrects most changes in base sequence, but about one in a million gets through. Some changes don't matter, but most are detrimental, and the cells that inherit the variants soon die out. Only a very few can provide advantages under one condition or another. Mutations result when the sequence of nucleic acid bases is not copied exactly, such as when an A is incorporated where there should be a G in the newly made strand of DNA. This error will then be propagated through the generations, unless it leads to the extinction of the line. Since each gene consists of hundreds of contiguous bases, random changes in a gene will result in random changes in the amino acid sequence of the protein it encodes. It would be like randomly changed letters in a word: "victory" might become "wictory" or "viktory" or "victori" or "cictory" or "oictory." The first three changes might still make some sense, but the last two are meaningless.

In 1943, Salvatore Luria and Max Delbruck carried out an experiment demonstrating that rare cells in a population of *Escherichia coli* were immune to a virus (Luria and Delbruck 1943). Most cells were killed by the virus, but a few (about one in a million) survived and gave rise to colonies even though they were surrounded by viruses. By diluting the population and growing small numbers of cells separately as subpopulations before adding the virus, they were able to show that the mutation in the DNA sequence appeared *before* it could do any good. In other words, before the virus was added. Only after being exposed to viruses did these rare mutations provide any advantage. It was later demonstrated that chemical mutagens increased the frequency of mutants that were subsequently shown to be resistant to viruses. The overall population was harmed by the treatment, but errors in copying the DNA in a small number of the cells made them virus resistant, and they gave rise to colonies in which all the cells were resistant. Natural selection, in the form of lethal viruses, gave the variants the ability to grow where their cousins could not.

The occurrence of rare mutants has been demonstrated in populations of many organisms, including bacteria, plants, and animals. They arise from random errors in replication of the sequence of nucleic acids. Most organisms use DNA as their hereditary material, but some viruses use RNA. Errors occur in replication of both DNA and RNA. In fact, the error rate in replicating RNA is higher as the result of having fewer proofreading mechanisms for copies of RNA than for DNA. RNA viruses use this property to evade immune responses of their host by randomly changing surface proteins, but pay a price in efficiency. Nevertheless, RNA viruses have not gone extinct and so should be considered successful parasites.

Random mutations generated a wide variety of different bacterial strains almost as soon as cells adapted to a new environment. Some of them acquired the ability to trap sunlight and to use photosynthesis to put the third, high-energy phosphate on ATP. Others adapted to chemically rich environments such as undersea vents. For the first few billion years, they were all anaerobic because there was little or no oxygen in the air. However, a by-product of photosynthesis is molecular oxygen, which gradually filled up the atmosphere. Oxygen is highly reactive and can quickly destroy many proteins unless they are specially selected for resistance. Random mutations in some bacteria allowed them to survive in an oxygen-rich environment and use it for more efficient metabolism.

The origin of eukaryotes can be traced to a chance association between an anaerobic archaebacterium and an aerobic bacterium (Margulis and Sagan 1995). It appears that an archaebacterium engulfed a bacterium, and they set up a stable relationship in which the bacterium provided efficient metabolism and the archaebacterium provided greater accuracy of replication. Thereafter, the DNA sequences encoding oxygen-resistant proteins were passed from the bacterium to the chromosomes of the archaebacterium, and they became forever dependent on each other. All the mitochondria now working in eukaryotes are derived from the original bacterium that was engulfed.

As the oxygen level increased, so did the size of some eukaryotic cells, and they started to feed on the smaller cells around them. Natural selection must have led to an arms race, where cells grew larger so as not to be eaten and others became even bigger to continue feeding. Present-day amoebae are thousands of times larger than the bacteria they feed on and several times as big as yeast cells, which they can also engulf. Natural selection acting on the random mutations that increased size led to some truly enormous cells found among the species of slime molds and certain algae. It also led to improvements in detecting the molecules released by prey, the ability to move more rapidly and organize the cytoskeleton, and the ability to engulf and digest food. To avoid these improved predators, some cells found an advantage in sticking together after cell division, forming large colonies that were difficult to attack. This led to the multicellular organisms we are familiar with, the plants and the animals.

The similarity of amino acid sequences in enzymes catalyzing the same reaction in even the most highly diverged eukaryotes makes it very clear that they are all related. Eukaryotes could have evolved separately several times, but one line predominated early on. They are all much more closely related to each other than they are to either ancestral line, the bacterial or the archaebacterial. The degree of similarity between comparable proteins can be used to establish the hierarchy of relatedness among all living eukaryotes. When averaged over thousands of comparable proteins, those in humans have sequences that are clearly more closely related to sequences in fish than to those of the fly *Drosophila* or the mosquito *Anopheles*. Humans and fish are both vertebrates, while flies and mosquitoes are invertebrates, so this is no big surprise. On the basis of sequence comparison, we are also more closely related to plants than we are to ciliates such as *Tetrahymena*, which is related to the malarial agent *Plasmodium*

(Song et al. 2005). Yeasts diverged from the line leading to animals more recently than amoebae or plants did, but their sequences changed more rapidly for a while, perhaps because they had come on land at a time when the ultraviolet light was far stronger than it is now and the irradiation increased the rate of mutation. In any case, primary sequence analyses clearly show the close relationship of all eukaryotic organisms.

CLEANING, WEEDING, AND HARVESTING

If the sink smells, we might pour in bleach and kill the germs. The strong chemical, sodium hypochlorite, rapidly kills bacteria and fungi and takes care of the problem. There may be billions of bacteria in the drain that are killed, but this does not concern us, because their life-forms are so different from ours that we do not relate to bacteria or fungi. Yet each of the cells that we kill was "doing something." They were exchanging material with the environment using highly evolved protein pumps embedded in their membranes and metabolizing components using enzymes that had been selected for billions of years. They replicated their DNA using the machinery that is common to all organisms, and they positioned copies in the daughter cells before dividing. They were certainly alive. However, microbial life is cheap.

Likewise, when we pull up weeds in a garden or field and throw them on a pile, we are terminating the life of these plants. We may share many of the fundamental genes with plants, but this does not stop us from keeping a tidy garden or a productive field. The weeds entered uninvited and have to be thrown out. Even when cutting down a full-grown tree, either to clear a field for planting or for firewood, we do not think about the life of the tree, since it had grown on its own with little or no help from us. To most people, plant life is cheap.

So, is it only nurtured life that we find precious? A farmer puts in long days preparing fields, planting a crop, and fertilizing the seedlings. As the crop grows, it may be protected from insects by insecticides sprayed at regular intervals. Some fields are planted with transgenic seed that has been carefully prepared to make a crystal toxin that kills nematodes which would otherwise reduce the yield. These are some of the most carefully nurtured seeds in the world, having benefited from years of intensive research and the most advanced genetic techniques. By the time the crop is harvested, it has profited from months of continuous care. The harvest is carried out with no thought to the life of the plants, only to the market price.

Many fruits and vegetables receive extraordinary individual care and nurturing. The Asian pear has been cultivated for centuries in China, Japan, and Korea and bred to produce delicious fruits. The pears are best when left to ripen on the tree, but some strains are fragile and subject to infestations. On certain farms each individual pear, while still on the tree, is wrapped in protective paper tied with a ribbon and left to ripen. Then, still wrapped, it is brought to market. The price alone indicates that it is a highly cared-for fruit. No one objects to eating a delicious pear, no matter how well nurtured.

So is it only animals that we find special? An angler pulls in salmon from the sea with little thought about the fact that they started life as eggs carefully buried by their mothers in a distant stream. The mother died shortly after, so the amount of nurturing was limited. But other salmon are farmed in ponds, where they are provided with food and fresh water, protected from predators, and generally watched over. They arrive on the dinner plate nonetheless. They may have been nurtured, but only in a commercial manner. Although fish have eyes and brains and are vertebrates like humans, they are still considered mostly as a source of food.

Until recently most humans were hunter-gatherers who lived off nuts, berries, fruits, and small animals. Wild animals were considered meat. Only in the last few thousand years have people domesticated sheep, goats, pigs, and cattle. They carefully raise and guard the animals before slaughtering them. In some societies, hunters and butchers follow a ritual of demonstrating respect to the animal before killing it, but mostly it is done with little thought about the life that is taken. The most extreme example of nurturing animals used as food may be Kobe beef, where each cow is hand raised, massaged daily, and fed beer before being brought to the slaughterhouse. Even in more normal cattle operations, cows suckle and care for their young for months, and cowboys often assist in birth, inoculate against disease, and provide plentiful food in feedlots. Then the cows and steers are turned into steaks and hamburger. Nurtured life may be precious, but we eat it every day.

HUMAN LIFE

Humans consider humans to be special. Everyone is aware of being human and wants to be treated specially. We certainly do not want to be considered someone else's food. As children we expect to be cared for, fed, clothed, and protected by our parents and relatives. As adults we strive for a society in which everyone is treated with respect and the weaker members are provided for. Human life is considered precious.

Birth is a defining moment but not a discontinuity in the life of the individual. The genetic makeup is determined when a particular egg is fertilized by a particular sperm. If all goes well, the resulting zygote divides to generate a ball of about a hundred cells, some of which give rise to the embryo while others participate in making the placenta. These extraembryonic cells have exactly the same genes as the fetus, but they are needed only during gestation. At birth, the placenta and umbilicus are discarded with little fanfare, even though the genetic inheritance of these organs is identical to that of the newborn. It is the life of the child that we care about, rather than the cells derived from the fertilized egg. And yet society is uncomfortable about using early embryonic cells for potential medical breakthroughs.

Almost everybody is charmed by and feels protective toward a human baby. Babies are beautiful and precious to us. They have enormous potential as human beings—we know they may grow up to be musicians, politicians, or scientists and may become caring parents themselves. We want to help them on their way. If lives are in danger, babies are usually the first ones helped into the lifeboats. Babies may not know how to walk, talk, or take care of themselves, but we know they have the ability to acquire these skills. Their developing consciousness is apparent in their wide-eyed attention, although it is not clear they have yet developed a sense of self. Over the next fifteen to twenty years, they will develop into young adults who are expected to take up their tasks and responsibilities in society. They will then be held responsible for their actions while afforded certain rights.

These same human beings, if convicted of certain capital crimes, may be executed. Serial murderers are rightly considered a menace to society and are often given the death penalty. In other societies, they are locked up in carefully guarded penitentiaries to separate them from innocent people. Countries that still have the death penalty use many arguments for putting murderers to death. Some argue that is it more humane than keeping a prisoner incarcerated for life. Others consider the financial savings in terminating the sentence relatively quickly. It is argued that the death penalty is a deterrent to potential criminals, but evidence that capital punishment significantly deters those prone to violent crimes is weak or nonexistent. Support for the death penalty comes in part from a need for revenge for the horrible crime. Many people feel that the criminal must pay the ultimate price for his or her completely unacceptable behavior. Those who do not support capital punishment argue that society is doing exactly what the murderer did, taking a life. They feel that all human life is precious and should be protected, even that of a vicious murderer.

In a series of decisions in the early 1970s the United States Supreme Court limited the rights of states to impose the death penalty, which it called a violation of the Bill of Rights, which protects us against "cruel and unusual punishment." A ten-year moratorium on executions followed. However, in 1976 the decision was overturned for specified cases. Gary Gilmore was executed by firing squad in Utah in 1977. Supreme Court Justice Harry A. Blackmun had voted for the reinstatement of capital punishment in 1976, but by 1994 he concluded, "I feel morally and intellectually obligated simply to concede that the death penalty experiment has failed. I no longer shall tinker with the machinery of death."

In October 2005, the European Human Rights Organisation pressured the United States and Japan to end capital punishment. "What is at stake is the most fundamental right to life and human dignity, and this is well worth a temporary drop of a few points in opinion polls. If the United States of America and Japan, as two leading democracies in the world, would abolish the death penalty, others would follow," said the group's secretary general, Terry Davis. Over the last twenty years, 117 countries have abolished the death penalty in law or in practice.

In 1986 the United States banned execution of the mentally retarded and, in 1988, limited the right of states to execute someone who was less than sixteen years old when the crime was committed. The following year the Supreme Court ruled that states were not prohibited from sentencing those aged sixteen or seventeen to death for capital crimes. The International Covenant on Civil and Political Rights prohibits capital punishment of anyone under the age of eighteen at the time of the crime. Although the United States ratified this covenant in 1992, it reserved the right to execute juvenile offenders. While usually professing compassion for children, judges and politicians in the United States argue that juveniles should be put to death for capital crimes. When does compassion for a child end and treatment as a guilty adult begin?

Guilt or innocence aside, people have been killing each other whenever tribes have clashed. Protecting homelands by killing enemies has been glorified and encouraged. Conquering new lands by killing and intimidating previous occupants has often been made to sound heroic, especially when told by the victors. Although there is an old religious commandment not to kill, it is usually interpreted as meaning one may not kill members of the tribe; others are fair game. It appears that the preciousness of life is contextual.

Modern societies encourage peaceful interchange among their citizens and between their citizens and foreigners. Diplomacy is usually thought to be far better than war. This does not seem to stop wars, either large or small, which break out frequently and lead to the death of soldiers and noncombatants. Efforts to civilize the process, such as treating prisoners decently and ultimately repatriating them rather than torturing and killing them, continue to be made. Indiscriminate bombing of civilian targets is discouraged but still occurs when it is deemed necessary.

By the age of twenty a person has been fed, clothed, educated, and taught to be polite. What value can we put on the individual?

This question has been asked by economists, lawyers, and politicians for many years, both in and out of court. Settlements in cases of wrongful death have varied from a thousand to 5 million U.S. dollars depending on the age of the individual, the affluence of the country in which the individual resided, and the immediate circumstances. When a U.S. Marine jet inadvertently hit the cables of an aerial tramway in Italy, killing all the occupants, the families of the victims were each given $2 million dollars as compensation. When sixty innocent Afghans at a wedding were killed as their village was strafed by mistake, the survivors were given two hundred dollars for each individual killed. If we use this measure of value, then an Italian is worth ten thousand times as much as an Afghan. Obviously, the value of life should not be decided by government bureaucrats.

Economists have made calculations based on future earnings models that consider human beings as machines generating streams of income. In this model, older people are considered less valuable than those in their prime. Retired people are worth nothing. Others have tried to determine what people are willing to pay to lower the probability of imminent death. The calculation gets complicated when the quality of life is brought into the equation. While these exercises may be important to corporations faced with class action suits for willful neglect, they demean life, reducing it to a number on a check. No amount of money can replace a loved one.

WHAT IS LIFE?

In the last century, research in biochemistry has discovered the pathways that convert sugars to fats and fats to amino acids and back and forth in an interconnected web that lets us eat what we want and still manufacture all the things we need in order to survive and grow. Plants trap sunlight and grow by taking up carbon dioxide from the air, and this happens not by magic but with specialized pigments attached to sophisticated molecular motors. We now know the proteins that make up the motors, and we know how they fit together to make ATP, which then powers all the other interconversions and reactions. Molecular biology has explained the universal mechanisms of heredity and how DNA can be copied and proofread so accurately that progeny are almost always exact copies of their ancestors. We know the code such that, when we know the sequence of bases in DNA, we know exactly which amino acids will be strung together to make this protein or that. The complexity held within a single cell is daunting but not beyond our ability to comprehend. Cell and molecular biologists who are willing to learn hundreds of reactions and general rules come to feel they know how a cell works. Mysteries still exist, but it is exciting to try to solve them.

The evidence is now solid that life has been present on earth for billions of years, and that natural selection has generated all the varied forms from descendants of the earliest cells. Very rarely, random errors were made while copying DNA, and most of these errors resulted in defective cells that soon died out. However, with billions and billions of cells all dividing and multiplying, now and then one of the variants was able to boldly go where none had gone before. Bacteria gave rise to eukaryotes with their carefully packaged chromosomes, and eukaryotes gave rise to plants and animals. While still a bacterial world, the surface became populated with multicellular organisms of all shapes and sizes.

We may consider a cell as a thing of beauty, but we do not see it as something that has to be protected at all costs. We reserve that feeling for our families, for those we have cared for and nurtured. We can admit that we are related to all living cells, but rather distantly, and we have no problem in eating a pear or cleaning the sink. People are more than just collections of cells. We value each other even as we compete with each other.

So, have we really defined life? Only in a materialistic, mechanistic sort of manner. There is so much more to life besides eating, digesting, and procreating that biochemistry and molecular biology have only scratched the surface. The real problems of life involve daily decisions about how best to get on and what is right or wrong. Some of these problems have been around forever, but others have come upon us so suddenly, as the result of recent advances in understanding cell biology and molecular genetics, that we are unprepared to respond. In the next few chapters, I consider biologically generated societal problems, explore the range of possible responses, and hold strongly felt values up to the light of recent biological understanding.

Technical improvements have made the process of in vitro fertilization an option for infertile couples who want to have children. Nevertheless, certain groups, including the Catholic Church, feel it is unnatural and oppose it on moral grounds. Islam permits in vitro fertilization but restricts it to using the sperm and eggs of a husband and wife. The technique puts the question of whether a fertilized egg is a person squarely on the Petri dish. Moreover, the process generates more embryos than are usually needed, raising the question of what should be done with them. At present, most are thrown out within a few months, but they could be used to generate embryonic stem cells with the potential to generate all the cell types of the adult body. Stem cell research holds out the possibility of being able to treat presently incurable diseases and intractable injuries. However, many countries either prohibit or severely restrict the generation of human embryonic stem cells on the grounds that a potential human life is destroyed in the process. The difference between potential and realized human life has to be clearly understood.

Therapeutic cloning, in which the nucleus from one of the patient's somatic, or body, cells is used to replace the nucleus of a human egg, is also contentious, because improvements in the technique might one day lead to personal clones that could be harvested for spare parts. However, embryos generated by somatic cell nuclear transfer (SCNT) have to be implanted in a womb to come to term, and no reputable doctor is proposing to do this at any time in the foreseeable future. SCNT techniques have been perfected in mice, and there are hundreds of clones running around in various labs. So cloning is not just a hypothetical question: it could probably be done for humans, but who would want to? I consider these and other questions in the next chapter.

Later chapters go into whether we should consider taking evolution into our own hands and design better organisms—pest-resistant crops, sheep that have useful drugs in their milk, humans free of hereditary diseases. Having sequenced the human genome, we can spot defective genes much more easily, but there are consequences to knowing too much, and this has to be taken into account. These are all aspects of human life.

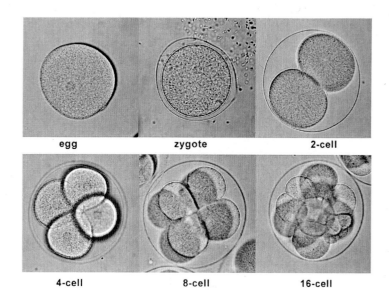

egg   zygote   2-cell

4-cell   8-cell   16-cell

TWO

HUMAN POTENTIAL

ONE SUMMER WHILE I WAS IN COLLEGE, I WORKED AT THE MARINE Biological Laboratory in Woods Hole, Massachusetts. Mostly I washed glassware and prepared solutions, but I had free run of the facilities and learned how to induce sea urchins to release eggs and sperm. The spiny animals were put into bowls of seawater and zapped with a low-voltage current. Females released huge numbers of yellowish eggs about the size of a pinhead, and males released sperm. When a dozen or so eggs were transferred to a shallow depression in a microscope slide and sperm were added, I could see that all the eggs were fertilized within a minute or so, because each one became surrounded by a raised fertilization membrane. An hour and a half later, each fertilized egg divided right down the middle to make two cells. An hour later both cells divided again, generating four cells, at almost

---

FIGURE 2.0 Fertilization and cleavage of a sea urchin egg. Four rounds of cell division produce sixteen cells that will continue to divide to make the embryo. Photo by Judith Cebra-Thomas.

exactly the same time. I was amazed. It looked as if a simple object, an egg, could spontaneously give rise to a more complex object, a four-celled embryo. How could a closed system become more complex in a regular fashion without the injection of energy? It was almost like a perpetual motion machine that breaks the Second Law of Thermodynamics. The answer was obvious when I remembered that eggs are filled with yolk, which can be metabolized during embryogenesis using the oxygen dissolved in the water. Each embryo comes with its own power plant that can drive processes like cell division and replication of the nuclei.

After being transferred into bowls, the embryos continued to divide, and by the next day each had become a hollow ball with hundreds of cells surrounding a central, fluid-filled space called a blastocoel. By the end of the week, the embryos had all developed into larvae with little arms held out around their mouths. They floated around for days before being discarded. Over the summer I fertilized many different batches of eggs, and they always followed exactly the same sequence of embryological steps characteristic of the species. My fascination led to a strong desire to understand the genetic basis of embryogenesis, and this interest has controlled my career ever since. Many years later I worked again at the Marine Biological Laboratory in Woods Hole and was still fascinated to watch the early steps in sea urchin life unfold. By then, I understood more about the molecular processes underlying the changes in shape and size of the cells, but still took an irrational pleasure in seeing them unfold. Nevertheless, after I had learned as much as I could, I poured each batch down the drain before it started to smell. There were lots of sea urchins available in the supply room, and each could give millions of eggs.

A sea urchin larva is surrounded by ectodermal cells that act as a skin; its gut is made of endodermal cells specialized for the uptake of nutrients. In between are mesodermal cells that make its skeleton (Gilbert 2006; Loomis 1986). All these cell types are derived from the single fertilized cell, and each carries the full complement of genes in its chromosomes. However, some of the genes expressed in ectodermal cells are never expressed in mesodermal cells. Likewise, there are endodermal-specific genes. Complex protein networks determine which genes are active in a given cell type. As development proceeds, genes are turned on and off in different cells so that they become specialized to carry out various functions during the larval stage.

When a sea urchin egg has divided into four cells, it is possible to tease them apart and allow each to develop separately. They will continue to divide and will form four hollow balls that go on to become four small larvae. Like the initial fertilized egg, cells at the four-cell stage can give rise to each of the cell types of the organism. However, if cells are isolated from an embryo that has gone on to make sixteen cells, they will give rise to only a restricted set of cell types. Having different patterns of gene expression during the first five to eight hours makes them different from each other and restricts their potential to make a complete sea urchin. The cells then are no longer able to give rise to all the tissues, as they were during the first few hours following fertilization. Only later, when they have become adults, will the ability to generate each of the cell types be reacquired by certain highly differentiated cells—the eggs.

IN VITRO FERTILIZATION

Human eggs are harder to come by than sea urchin eggs. A woman usually releases only a single egg into the fallopian tube each month, where it can be fertilized naturally. However, about 10 percent of the women of reproductive age in the United

States are infertile, often as the result of blocked or damaged fallopian tubes. For the last twenty-five years, women who wanted to become pregnant, but who could not do so naturally, went to clinics for help. The desire to have children is very strong in many people, and they will make great efforts to become pregnant. Fertility clinics are able to collect eggs directly from the ovaries and fertilize them with sperm in glass containers (thus the name "in vitro fertilization"). A few days later the fertilized eggs are inspected, and those that have divided to give rise to blastocysts with eight or more cells are positioned in the woman's uterus with a catheter. Sometimes in vitro fertilization clinics like to wait for a few more cell divisions before transferring the cluster to the patient. At the transfer stage the blastocysts are not really embryos, since most of the cells will form extraembryonic tissue, so they are referred to as pre-embryos.

To increase the chances of success, a dozen or so eggs are fertilized at the same time. They are collected after the patient is treated with hormones to induce simultaneous maturation of multiple eggs. Guided by ultrasound scans, the doctor collects the eggs with a needle inserted through the peritoneal wall. Typically five to fifteen eggs are collected and fertilized with a hundred thousand motile sperm. If necessary, sperm can be injected into individual eggs. A few days later, two to four pre-embryos are placed in the woman's uterus, since not all will implant successfully. Excess pre-embryos can be preserved by freezing. If the first cycle does not result in pregnancy, pre-embryos can be thawed for use in a second cycle. The patient owns the excess pre-embryos and can chose to use, donate, or discard them. Success is determined by a standard pregnancy test after a few weeks and confirmed by sonograms of the womb after a month. The rest is left up to nature to complete.

The techniques of in vitro fertilization have improved dramatically since the first "test-tube baby" was born in England in 1978. The process is safe, and no ill effects on either the mothers or the offspring have been encountered in thirty years. The chances of success now approach 50 percent at some clinics, and their freezers are full of unneeded pre-embryos. The eggs that generated two of my grandchildren were fertilized in vitro.

Nevertheless, the Catholic Church condemns in vitro fertilization as an evil act that is a threat to the unity and stability of the family. In 2004 Pope John Paul II voiced concern that the procedure allowed a human embryo to be "treated as a product of technology and not as a gift of God." Such fertilization was felt to establish the "domination of technology over the origin or destiny of the person" (Shea 2006). Therefore, the church concludes, in vitro fertilization is morally unacceptable, and the creation of any single-cell human organism by any technique should be banned. The problem comes from considering a fertilized egg to be equivalent to an organism. Humans are not single-celled organisms. Infertile Catholic couples have been advised to pray to Saint Anthony and hope to become pregnant by the God-given way.

The Islamic view of in vitro fertilization is that it is permissible as long as the sperm and egg are from a husband and wife who are legally married. Even when the husband's sperm count is too low for fertilization, use of donated sperm is not permissible. Likewise, even when the wife's eggs are defective, donated eggs cannot be used. Giving the embryo to another woman to carry as a surrogate mother is also not permissible. The Islamic Fiqh Council ruled that such practices would confuse parentage and inheritance rights. Sometimes the desire for having children overcomes financial worries.

EMBRYONIC STEM CELLS

There are hundreds of thousands of pre-embryos now stored at fertility clinics that ultimately will be discarded. Some of them could be put to use as the source of stem cells for medical research or even therapy once some basic developmental problems are better understood.

How are embryonic stem cell lines established that can grow indefinitely in the laboratory? As is often the case, techniques for making human embryonic stem cells (ES cells) were first worked out using the mouse, a major model for mammalian fertilization and embryogenesis. A few days after fertilization, a group of cells is formed within the blastocoel that is called the inner cell mass. The embryo develops from these cells, while the surrounding cells form support tissues such as the embryonic contribution to the placenta. When cells are dissociated from the outer wall of a blastocyst, they grow but give rise only to extraembryonic cell types. Cells from the inner cell mass, however, differentiate into a wide variety of cell types. They grow rapidly when deposited in a Petri dish on top of a feeder layer of skin cells, which add growth factors and hormones to the nutrient medium.

As long as the embryonic cells are kept uncrowded and well fed, they produce populations of identical cells that show no evidence of differentiation. They can proliferate through hundreds of cell generations, producing enough cells for almost any conceivable experiment. Moreover, individual cells can be isolated and grown clonally to establish lines in which all the cells come from a single cell. When they have grown so much that they crowd the dish or are shifted to a more limiting medium, they differentiate into a wide variety of cell types. When mouse ES cells are injected into mice, they form teratomas, which contain cells characteristic of gut, bone, cartilage, smooth and striated muscle, neural epithelium, ganglia, and stratified epithelium. These are representatives from the three major embryonic cell layers: endoderm, mesoderm, and ectoderm. Cell lines that are capable of unlimited proliferation, as well as able to differentiate into all the cell types when incorporated into an embryo, are considered ES lines.

Just as the cells of a four-celled sea urchin embryo are able to generate all the cell types of the organism, some of the cells in a human pre-embryo can differentiate into almost any cell type, even after being grown in the laboratory for weeks. Researchers must use human cells to learn how to guide the developmental pathways followed by stem cells if the resulting knowledge is to be used for medical treatments. However, society is not at ease with stem cell studies that require the destruction of pre-embryos, even unneeded ones that will be discarded regardless (Gilbert, Tyler, and Zackin 2005). The National Institutes of Health, the large federal funding organization, has limited all government support in the United States for embryonic stem cell research, usually called ES work, to studies done on lines created before August 9, 2001. There are only a half dozen or so embryonic stem cell lines presently used in NIH-funded studies, and many show signs of obvious defects. While most problems could be avoided by the establishment of new stem cell lines, such work can be carried out in the United States only with private financing. The governmental authorities in the United States and several other developed countries do not want to be seen supporting activities in which potential human life is terminated, and they fear that some human eggs may be fertilized expressly for the production of ES cells.

In mice it is possible to develop a breeding colony from the ES cells of a single embryo. Cells from the inner cell mass are injected into the blastocoel of another embryo, where many of them join the inner cell mass. These combined cells go on to develop into all the organs, including those that give rise to eggs or sperm. If the strain of mouse from which the ES line is derived is genetically distinct from the strain of mouse used to make the host embryo, the resulting offspring will be made up of cells from both strains. It is a chimeric mouse, with some cells derived from the ES line and others from the host blastocyst. Some of the sperm produced by chimeric males carry the genes from the ES line, and others carry the genes from the host strain. By mating chimeric mice together or backcrossing to one of the parental strains, animals derived only from the ES line can be established. Most strains generated in this manner are healthy and thrive, thereby demonstrating that the ES line is able to generate all the cell types found in an adult. Although ES cell lines grow as apparently undifferentiated cells in the laboratory, they can produce all vital organs when incorporated into an embryo. A large number of mouse strains have been generated in this way to better understand the roles of specific genes.

Work with human ES cell lines followed the same course, but researchers stopped short of injecting them into the blastocoel of a human embryo since there were serious ethical concerns about creating a chimeric individual (Thomson et al. 1998). However, by injecting human ES cells into immunodeficient mice it was possible to show that these cells can create endodermal, mesodermal, and ectodermal tissues. The procedure uses cleavage stage pre-embryos produced by in vitro fertilization that are donated by individuals after informed consent. The pre-embryos are cultured until cells from the inner cell mass can be collected and deposited on a layer of mouse skin cells. After about a week, many of the cultures will have attached and spread, and dissociated cells can then be moved to fresh layers of skin cells. In a study published in 1998, James Thomson and colleagues showed that cell lines produced in this way all had high levels of an enzyme, telomerase. This is characteristic of immortal cell lines, which can grow continuously in the lab. They also found proteins and sugar groups on the surface of the ES cell lines that are usually found only on cells of the inner cell mass. When the populations were allowed to overgrow in a culture until the cells piled up on each other, they started to differentiate spontaneously into specialized cell types.

The ability to differentiate into organs could be demonstrated only by injecting the cell lines into mice, so that they formed teratomas. These chaotic masses of cells occur when stem cells give rise to tumors in the testes of men or the ovaries of women. They can be benign or malignant. Some teratomas have well-developed teeth, hair, fat, and nervous tissue. Those formed by human ES cells following injection into mice develop in the same manner and generate a variety of tissues. They do not appear to be any different from mouse ES cell lines, except that they carry the genetic complement of humans.

The establishment of ES cells from in vitro fertilized pre-embryos is now a mature technique with a high degree of success. However, it takes some expertise to generate and maintain the lines. Few laboratories are able to carry out such work without federal funding for the necessary expenses. The ban on support for all work with human ES cell lines generated after August 2001 has limited embryonic stem cell studies in the United States. As a result, progress in understanding the mechanisms that guide cells down the developmental pathways to specific cell types and the generation of functional tissues has been slowed. The ban was put in place in response to abortion opponents who did not want the federal government to condone or facilitate anything that resulted in the destruction of an embryo. Removing the inner cell mass clearly precludes a pre-embryo's ability to ever give rise to an embryo. However, the pre-embryo would have been discarded in any case.

A recent procedure has shown that removing one cell at the eight-cell stage leaves the remaining seven cells able to continue embryogenesis. The single cell can be grown in a Petri dish and, in some cases, will give rise to an ES cell line. The remaining seven-celled pre-embryo can then be implanted in hopes that it will develop to term. The success rate with seven-celled pre-embryos is not significantly different from that with eight-celled pre-embryos. For a number of years, fertility clinics have offered their patients the option of preimplantation genetic diagnosis if there is reason to think that some fertilized eggs might carry defective genes. In this procedure a cell is collected at the eight-cell stage following in vitro fertilization, and DNA tests are run for any of 150 common genetic defects. If the tests show it to be free of these genetic diseases, the blastocyst can be implanted. The number of successful births following preimplantation genetic diagnosis attests to the safety of the procedure. The single-cell method for generating ES cells is no different, but it does not work every time. In one study a total of sixteen human pre-embryos generated from in vitro fertilized eggs that were tagged for disposal had to be used to make two ES lines. Efforts are now under way to improve the success rate by fine-tuning the growth conditions for the cells. Establishing ES cell lines while leaving the pre-embryo viable removes many of the ethical problems associated with destroying embryos, but some ethicists are troubled by the question of whether the extracted cell itself has the potential for life. And some are against in vitro fertilization itself. It is unlikely that the technique will settle the debate.

Perhaps someday every child born after in vitro fertilization will have its own ES cell line stored for the future when it may be needed as a genetically identical source of missing or diseased tissues. Many diseases, such as Parkinson's disease and juvenile-onset diabetes, result from death or dysfunction of a single or a few cell types. Replacement of those cells with healthy ones derived from the individual's ES cell line could offer a lifetime cure. However, such treatment would be possible only after a considerable improvement in our understanding of the basic developmental mechanisms that direct ES cells to desired lineages. Uncontrolled growth that could result in tumors would have to be reliably avoided. Many biologists are interested in these problems, but they are constrained from working on them by present funding restrictions.

Transgenic mice that glow in the dark have been derived from ES cells. This may sound like a party trick, but it allows the fate of specific cell types to be followed in whole animals. The fluorescence comes from a jellyfish protein that gives off green light when excited by ultraviolet light. The gene encoding this protein, called GFP, can be engineered in the lab to be controlled by the regulatory region of a normal mouse gene expressed in only a single tissue. ES cells that have incorporated this engineered DNA can be selected while growing in the undifferentiated state and then introduced into the blastocoel of embryos. Some of the transformed cells will enter the inner cell mass and participate in the formation of the embryo. If the GFP gene is engineered to be controlled by the regulatory region of a gene active only in midbrain cells, then only the midbrain shows fluorescence (Zhao et al. 2004). These transgenic mice provide a valuable tool for better understanding the cause and possible cure of Parkinson's disease, which results from impairment of cells in the substantia nigra, the part of the midbrain responsible for production of the neuronal signal dopamine.

ES cells with the engineered DNA that are grown as undifferentiated populations in the lab do not express this GFP construct or show fluorescence, but when they start to differentiate, some of them glow green. If this cell line is further modified so that the cells also express high levels of a particular transcription factor that stimulates gene expression in the midbrain, SoxB, the proportion of green-glowing differentiating cells is greatly increased. Such control of lineage decisions will be of enormous value for therapeutic treatment in humans. But we cannot assume that what is true for the mouse is always true for humans. These experiments will have to be repeated with human ES cells. Molecular markers for specific cell types can lead the way to understanding how to preferentially direct ES cells to one cell type to the exclusion of all others.

THERAPEUTIC CLONING

Another approach to generating ES cells that could be used in tissue repair and transplantation medicine is known as "therapeutic cloning." It involves removing the nucleus from an egg and replacing it with the nucleus from adult tissue of an individual. The resulting egg is then induced to divide and form a blastocyst. ES cells can then be recovered from the inner cell mass and grown in the laboratory. The technique is also called "somatic cell nuclear transfer" (SCNT) to avoid the loaded word *cloning*. Both terms are correct. The nucleus used to replace the egg nucleus comes from a somatic, or body, cell, and each of the cells in the pre-embryo carries an exact replica of the transplanted nucleus—it is a clone of the individual who donated the adult tissue. It is thought that the nucleus of the somatic cell is reprogrammed when it enters the egg environment, turning off the genes expressed in the adult tissue and activating embryonic ones. Other processes have to be reversed as well before a nucleus is capable of starting all over again and making a new individual. The pre-embryos are allowed to develop to the blastocyst stage when their inner cell mass is removed to generate embryonic stem cells. No new individuals are made in therapeutic cloning, only ES cell lines.

Like almost all advances in medicine, approaches to therapeutic cloning followed studies with model organisms. More than fifty years ago, Robert Briggs and Thomas King (1952) found that they could take the nucleus out of a frog egg and replace it with a nucleus from an embryo with about a hundred cells. The transplanted nucleus replicated and the cell divided. Over half of the transplants developed into feeding tadpoles. They ran into troubles when nuclei were taken from cells at later stages of development. Almost none of the transplants made it to the tadpole stage. Did something happen during embryogenesis that made it impossible for a nucleus of a cell destined to make only one small part of a tadpole revert to the state from which it could direct any part of embryogenesis? About ten years later John Gurdon, working with eggs of a South African clawed frog, had better luck and could consistently get tadpoles from eggs carrying nuclei transplanted from late-stage embryos (Gilbert 2006). He was even able to get a few enucleated eggs to develop to the tadpole stage when nuclei from fully differentiated larval gut cells were used. However, most of the tadpoles died and never went through metamorphosis to give rise to fertile adults. While it seems that individual adult amphibian cells carry all the genes necessary to make the hundreds of different cell types of a tadpole, they don't work exactly as they should, and the tadpoles sicken and die. We still do not know why.

These early successes led to the development of techniques for nuclear transfer to mouse eggs and their development to the blastocyst stage, from which ES cells could be recovered (Wakayama et al. 2001; Kishigami et al. 2006). These clonal ES cells were grown in the laboratory to expand the population; some were then implanted into the blastocoel of a pre-embryo, where they joined the inner cell mass and went on to participate in the formation of all the tissues and organs. These ES cells could differentiate into nerve cells in the lab and into sperm when they were incorporated into chimeric mice. By 2003 it appeared that there were few barriers to generating ES cells following nuclear transfer. However, despite heroic attempts, so far no human ES cells have been generated following nuclear transfer.

In a flurry of papers published in 2004 and 2005, a large Korean team of veterinary scientists led by Dr. Woo Suk Hwang reported that they had succeeded in generating patient-specific ES cell lines by injection of nuclei into human eggs from which the chromosomes had been removed. Within six months it was found that the results had been fabricated and that there was no evidence for any ES cells derived from nuclear transplant, although at least 2,221 eggs from 119 women had been used. The team was highly skilled in nuclear transplantation for the generation of ES cells and was the first to clone a dog. Snuppy, the cloned Afghan hound, turned out to be a real clone, but the human ES cells all turned out to be derived from blastocysts that had been made by in vitro fertilization of human eggs by human sperm. They were not patient specific. Further investigation into the Hwang operations uncovered cases where women had been coerced or paid for their eggs, which was strictly against the ethical rules. No one wants to be an egg mill for some research group. Other cases of misuse of funds turned up in the investigations of one of the biggest cases of scientific fraud in the recent past. No one knows why Hwang seemed to have gone crazy in an attempt to be the first and best at generating patient-specific ES cells, but the Korean courts of justice are now in the process of trying to sort out the mess.

ES cell lines generated from patients suffering from particular diseases would greatly advance laboratory studies of the progression of those diseases, which might lead to ways to treat the underlying genetic defects in patients. Cells could be recovered from individuals with early-onset diabetes, Lesch-Nyan's disease, Parkinson's disease, cystic fibrosis, Alzheimer's disease, Lou Gehrig's disease (amyotropic lateral sclerosis), cancer, and other diseases. After fusion with an enucleated egg, the inner cell masses from resulting blastocysts could be cultured to give rise to ES cells carrying the genetic defects that trigger the diseases. Examining how these cells develop in vitro could reveal much about basic biology and the genetics of the diseases. Techniques have already been developed to direct populations of ES cells growing in Petri dishes to differentiate into pancreatic cells, nerve cells, muscle cells, and lung epithelial cells, which could be used to explore therapeutic treatments. Such cells could provide critical experimental material for drug screening and gene therapy that would more accurately reflect the diseases leading to diabetes, mental retardation, dopamine production, neuromuscular degeneration, mental degeneration, and cancer. Some of these diseases result from mutations in several different genes, and so it would be important to establish a bank of ES cell lines from a wide range of donors.

The day may come when this technology can be used to treat AIDS patients whose immune systems are impaired by HIV viruses. A somatic cell from a patient could be fused with an enucleated egg to generate a blastocyst from which ES cells can be recovered. After the ES cells are grown up to generate a large population, rare cells could be selected that fail to make the receptors which the virus uses to attach to the cells. There are a variety of ways this could be done using molecular biology and direct selection. HIV-resistant cells would then have to be coaxed into differentiating into immune cells before being injected back into the patient. While this has not yet been attempted, there are several promising approaches that could be followed. AIDS is such a dreadful disease that almost anything is worth trying that might work to defend against it. However, establishing personalized ES lines for every HIV-infected patient is impossible considering the huge number of AIDS cases. But it could work for a few.

Is there anything intrinsically wrong with establishing patient specific ES cell lines? Even though the Korean attempts failed, there was no loss of life. The donors were fully informed that the eggs they provided would be used for therapeutic cloning research and expressly told that the eggs would not be used for reproductive cloning. Somatic cells were fused with eggs from which the nuclei had previously been removed and the resulting cells coaxed into dividing in the absence of sperm by manipulating the calcium concentration. Although the women may have been subject to some discomfort, they were not harmed. Their eggs gave no useful results but were unlikely to be fertilized in the natural course of events. A human cell cannot be considered a life.

Many pro-life supporters believe that a person comes into existence at conception—when a sperm fuses with an egg. Some of them believe that SCNT is sufficiently similar to conception that a person comes into existence during therapeutic cloning. But genetically, this person already exists in the individual who donated the somatic cell. Nonetheless, pro-life advocates consider the destruction of an embryo to be murder. An embryo, regardless of its stage of development, is still potentially a human being in the making, they argue. However, even eggs that develop to the blastocyst stage after SCNT might not develop further, even if implanted in the womb of a receptive woman. They are only a ball of a few hundred cells with an inner cell mass. The potential might be there to generate a human, but it is only potential. Blastocysts are laboratory subjects given the impersonal but meticulous care that any cell line or tissue is given in the lab. When the inner cell mass is destroyed to establish an ES cell line, a life is not lost and the patient may be cured.

In ten years or so, therapeutic cloning might save countless lives and increase the quality of life for many others. If ES cells were available for those with diabetes, Parkinson's disease, or cystic fibrosis, they might benefit from receiving healthy cells that could make insulin, dopamine, or the chloride channel CFTR that is defective in patients with cystic fibrosis. Ideally, these ES cells would repopulate the pancreatic islets, the brain, or the lungs of the patients. Those suffering from liver disease might benefit from receiving ES cells generated from one of their own somatic cells that had been treated in vitro to differentiate into liver cells. The list of potential problems that might be treated with genetically matched cell types goes on and on.

The Christopher and Dana Reeve Foundation has been committed to finding treatments and cures for spinal cord injuries ever since Christopher Reeve, an actor who played Superman, was paralyzed after a fall from horseback. The foundation actively champions studies on ES cells for replacement or transplantation therapies of diseases and disorders such as spinal cord injury, diabetes, cancer, and Parkinson's disease. They argue that ES cell research could provide a deeper understanding of cell differentiation and development, possibly leading to insight on how to help the body heal itself. They suggest that the federal government should fund work on newly generated ES cell lines from some of the four hundred thousand surplus pre-embryos now stored in the United States that will otherwise be discarded. The foundation points out that embryonic stem cells have the potential to become any cell in the body. They encourage and fund work that may bring this potential closer to being realized at the bedside. Reeve himself hoped to be treated with ES cells to help regenerate his spinal cord. Unfortunately, he died in 2004.

If developmental biologists learn how to generate whole organs from SCNT-derived ES cells, it is conceivable that these new organs could be transplanted to replace diseased or destroyed organs. It is first necessary to understand the mix of cell types in such organs as the liver, kidney, and heart, as well as to encourage vascularization and innervation so that the transplanted organ can have a supply of blood and be able to respond to neural signals when put back into the patient. This is a tall order, but many organs are at least partially self-regulating under the proper conditions, and it is conceivable that someday it may be possible to generate functioning organs. Just not tomorrow. But think of the advantages of having a transplant with no danger of rejection because the cells are the patient's own. Since a new organ could be grown when needed, there would be no need to wait for an appropriate donor to die.

SOMATIC STEM CELLS

Throughout life, many tissues need to be continuously supplied with new cells generated by a specialized population of stem cells. Skin cells are full of keratin and never divide; they are half dead when they take up their role on the surface of the body. After two to four weeks, they die and are sloughed off and replaced by new skin cells generated from stem cells. These cells are constantly dividing so that they can both replace themselves as stem cells and generate precursors to epithelial cells. The new skin cells start to fill with keratin and move to the surface. You get a new skin every month or so. It is still you, but you are covered in new cells.

Likewise, the cells lining your gut are subject to the harsh chemical environment that digests almost everything that goes into your stomach. Cells have to project into the intestine so that they can take up nutrients, but they too are slowly digested and have to be constantly replaced. Although fully differentiated intestinal epithelial cells that extend into the gut do not divide, there are a few stem cells scattered along the intestine that divide every twelve hours or so. On average, half the stem cell progeny continue dividing, while the other half gradually become columnar shaped and put out thin extensions to increase their surface area on the side that points into the gut. As these cells differentiate, they replace the dead cells as they are shed. Because stem cells are maintained throughout the lifetime of the individual, they must be able to adjust to many different conditions and maintain a steady population through thousands of cell divisions. Under certain medical circumstances, it might be useful to manipulate the proportion of cells destined for terminal differentiation to give a burst of new epithelial cells. However, the signals regulating these stem cells are presently unknown. We have to think of other ways to repair intestinal injury. If intestinal stem cells could be isolated and grown in the laboratory, they could be transplanted back into patients, where they might replenish the intestinal epithelium. Unfortunately, intestinal stem cells cannot be identified morphologically or distinguished from surrounding epithelial cells by any known set of markers. For now we have to let the body take care of itself.

Blood cells too have a hard life as they course through capillaries, bumping into obstacles and being forced through small openings. The average red blood cell lasts only 120 days. Since there are billions of blood cells in every liter of blood, there has to be massive replenishment. In mammals the stem cells that give rise to red blood cells and circulating immune cells are found in the marrow of long bones. In this protected place they renew themselves as well as produce the precursors of red and white blood cells. For more than ten years, patients with anemia, leukemia, lymphomas, and diseases of the immune system have received bone marrow transplants to provide them with new blood stem cells after their own have been destroyed by radiation treatment, chemotherapy, or a specific disease. Every effort is made to match the donor and the recipient, first looking to close relatives, but there are often complicating problems when the grafted immune cells of the donor attack skin and liver cells of the host. Strangely enough, those with acute graft-versus-host disease can be treated with other human stem cells, which suppress adverse immunological responses and help the damaged tissue to heal. However, there are complications with this treatment as well, since these stem cells can also be recognized as foreign. It would be far better to use stem cells isolated from the patient and expanded by growth in the laboratory. However, just as it has not been possible to recognize and isolate intestinal stem cells, it has not been possible to establish pure lines of blood stem cells in the laboratory.

Fat appears to be a source of stem cells that can repopulate a variety of tissues, including heart and breast tissues. There is a ready source of fat tissue, as hundreds of thousands of Americans each year undergo liposuction to improve their looks. However, direct evidence for stem cells in fat tissue is quite thin. Rigorous experimental work with cloned cell lines has not been carried out yet. It is possible that injected fat cells somehow trigger repair mechanisms in surrounding cells rather than participating themselves.

Fat stem cell therapy has not been clinically tested or approved either in the United States or in Europe, but this has not stopped wealthy Russians from going to beauty clinics in Moscow for injections with fat cells. There is anecdotal evidence that facial injections lead to the replacement of fat and muscle tissue that is gradually lost during aging, thereby letting the recipient avoid the cadaverous appearance of the old. Some believe it has helped them to appear younger and feel more fit. They continue to pay many rubles to have fat tissue moved from their hips to their faces. However, one businessman found that he developed small skin tumors at the sites of injection. But this did not seem to lead to his loss of confidence in the technique, since he apparently had the tumors removed, went to another clinic, and had the procedure repeated. Clearly, some people are convinced that adult stem cells can be manipulated in sophisticated ways. However, all available data indicate that adult stem cells have very limited flexibility and can be dangerous if mishandled. Much more must be learned about the mechanisms that control proliferation versus differentiation, as well as about the processes that limit a given adult stem cell line to a restricted set of terminally differentiated cells, before somatic stem cells can enter general medical practice.

The life of an individual should never be taken lightly. But a few cells or even a mature egg is not a life. Cells are generated continuously, and many die naturally. Those in culture are no different. If the nucleus of an egg is removed and replaced with that of a somatic cell, it is still just a cell. Once the somatic nucleus has been reprogrammed, the egg can be activated and will often develop at least to the blastocyst stage. Removing the inner cell mass destroys the blastocyst but allows ES cells to be cultured in the laboratory. These cells would be perfect for therapeutic treatment of the donor since they carry the identical genetic complement. Moreover, they could serve as a source of experimental material. Although each blastocyst has the potential to give rise to an individual, at this stage they have no self-awareness or consciousness—they don't even have nerve cells. Human blastocysts have the potential to become human, but only after they have developed in the womb, formed a central nervous system, and developed the ability to breathe on their own when they are born.

Only a few years ago, many people thought it would be impossible to generate cloned mammals from somatic nuclei. However, improvements in transplantation and culture conditions, along with heroic efforts involving hundreds of eggs, have recently generated healthy, fertile, cloned mice, sheep, cows, dogs, and cats. Moreover, ES cells have been grown in the laboratory and genetically engineered to express genes from other species before being used as a source of nuclei to generate new mice and sheep. Such experiments have not yet been successfully carried out with human ES cells, but when the techniques are worked out, they may open Pandora's box. These matters are discussed in the following chapter. They raise all sorts of moral dilemmas that will have to be confronted. Rationally deciding how to proceed will require some deep thinking about who we are and what we want to become. The biological basis for individuality, for genetic control of behavior, thinking, and deciding, is covered in later chapters.

THREE

ENGINEERED LIFE

WHILE STUDYING CONTROL OF GENE EXPRESSION IN GRADUATE school, I learned how to genetically manipulate the bacterium *Escherichia coli*. I focused on the genes that allow the cells to grow on lactose, the sugar found in milk. These bacteria only express the genes for lactose metabolism when there is lactose in the environment, and they repress transcription of this set of genes when there is no lactose around. This seems sensible, but what if there is an even better sugar in the medium along with lactose? In the presence of both glucose and lactose, the bacteria first use the glucose and only later use the lactose. This level of sophistication is the result of a positive control mechanism required for transcription of the genes for lactose metabolism. When bacteria metabolize glucose, this control mechanism is blocked. We were able to show that the two control mechanisms act independently of each other, and that feedback loops amplify the effects. We

FIGURE 3.0 Dolly. Courtesy of the Roslin Institute, United Kingdom.

would make mutations in the genes and show how they affected the different components responsible for the exquisite control seen in the normal (wild type) cells. We could test our models by swapping the mutations around in different strains and built up a convincing story.

Although I carried out these experiments long before anyone had thought it possible to put genes in DNA elements that could be moved around, I could make genetic crosses and select for almost anything I wanted. I was able to create pure populations of strains with genetic complements that had never before existed. This is now routine in molecular biology, but it opened my eyes to what could be done. We were directing the evolution of the strains so that they could adapt to novel conditions. It was an exciting time, but it made me think that, if this was possible with bacteria, it might someday be possible with humans. It worried me that we might not have considered the disadvantages along with the advantages of controlling our own evolution.

When I brought up this concern with one of my advisors, Salva Luria, he promptly replied, "Not in our lifetimes, Bill." Salva was a highly respected microbial geneticist and a thoughtful man whose judgment I highly respected. I did not argue with him at the time, but as the years went on and genetic engineering developed to the point that it was possible to put specific genes into bacteria as well as mammalian cells, I pointed out to him that we were getting closer to the day when it might be possible to make genetically modified humans. I argued that the possibilities should be widely discussed before we embarked on what might be a one-way trip. Salva never thought it a serious problem for society. He worked hard on other problems, such as nuclear weapons testing, and was an outspoken critic of the American involvement in Vietnam. He participated in discussions about limits on the use of genetic engineering but never considered directed evolution imminent.

GENE THERAPY

Soon after it was discovered that specific pieces of DNA and even whole genes could be stitched into bacteria and recovered in large quantities after the bacteria had grown into large populations, doctors started considering the possibilities of curing certain diseases by replacing faulty genes with healthy genes. Molecular medicine seemed like a whole new way to treat the sick. These doctors were not thinking about directing evolution but only of helping those who were deathly ill. The treatment would be for the individual, not the progeny. By the turn of the twenty-first century, progress in sequencing the human genome led to an increase in the number of single gene defects known to result in hereditary diseases, adding further pressure to treat them by gene therapy.

For years doctors had known about the genetic defects causing the illnesses of some patients without being able to do anything about it. For instance, a few babies are born every year without a functional gene for the enzyme adenosine deaminase (ADA). They suffer from severe immunodeficiency and have to be completely isolated from the world around them to protect them from lethal infections. They live their lives in plastic bubbles. If only it were possible to give them a good copy of the ADA gene, they could get out and live a more enjoyable, rewarding life. It did not seem so difficult to harness the techniques that had been developed for transforming genes into cells to deliver genes to patients. It has turned out to be exceedingly difficult.

Isolating the human ADA gene was relatively easy. When introduced into cells, it functions well and produces the expected enzymatic activity. The problem was how to get the gene into patients so that it stably integrates into the chromosomes of the appropriate cells and gives long-lasting protection. A variety of approaches were explored using mice as experimental organisms. Some of the most promising techniques to come out of this work use viruses to deliver the gene to billions of cells, where it can be integrated in the nucleus along with the virus. A viral infection sounds like something we want to avoid, but the viral vectors that are used have been crippled so that they do not replicate within the body. They just enter the cells and lie there, unless they integrate into the chromosomes, in which case they are replicated along with any gene they carried in with them. Viruses that cause the common cold, adenovirus and adeno-associated virus, are popular vectors, as are retroviruses that go in as RNA but are copied into DNA, which inserts itself into the chromosomes. Choice of vector is based on efficiency, safety, and the frequency of stable integration. One of the drawbacks to using adeno-associated virus-based vectors is that they predominantly integrate into dividing cells, and the vast majority of the cells in the body are not dividing at a given time.

Other approaches involve mixing the DNA of the therapeutic gene with fatty acids and shaking them together until they form small droplets. These tiny liposome droplets are thousands of times smaller than cells and can be injected into the bloodstream, where they circulate and can even cross the barrier to the brain. At times the droplets will fuse with the surface membranes of cells, which are themselves largely made up of fatty acids. Once inside the cell, a few of the millions of copies of the gene find their way to the nucleus before being broken down. Once in the nucleus, one of the copies of the healthy gene might stably integrate into a chromosome and start producing the missing enzyme.

Part of the problem with gene therapy is that there is no way to know where in the chromosome the gene will integrate. If it is in the middle of another gene already in the chromosome, it will disrupt that gene and possibly lead to other severe consequences. This possible drawback was thought to be minor, since only a small percentage of the DNA in human chromosomes codes for proteins. The rest appears to be dispensable "junk" left over from failed duplications. Integrating in the dispensable regions of chromosomes would be of no consequence, and the transgene could provide the missing protein. Integration turned out to be a more serious problem than originally envisioned, since some viral vectors integrate preferentially into genes that control cell growth. When this happens, the cells can become cancerous and threaten the life of the patient in a new way. In 2003, ten children with severe immunodeficiency were experimentally treated in France with ADA carried on retroviruses. At first, they all seemed fine and were apparently recovering their immune responses, when three were found to have developed leukemia as a result of the mutagenic effects of inserting the gene. The sick patients were treated with chemotherapy, but one died. Although the gene therapy had been successful, further trials were immediately halted until the problem could be understood and avoided.

Attenuated adenoviruses appear to be safer vectors, but their efficiency is lower. Protocols developed with mice have been adapted to humans in clinical trials and some partial successes achieved. However, in 1999 Jesse Gelsinger died after being treated with adeno-mediated gene therapy. He had suffered from a lack of ornithine decarboxylase and was given doses of adenovirus carrying the missing gene. However, the amount of adenovirus turned out to be much too high, and it overwhelmed his ability to tolerate the virus. These trials too were stopped until the protocol could be perfected.

Throughout these early steps toward adding gene therapy to the medical arsenal, federal and institutional biosafety committees grappled with the ethical implications. The first directive to all doctors is to do no harm. Although the procedures had been demonstrated as safe in trials with mice and rats in the laboratory, they had to be shown to be safe for the patient and the hospital workers. Construction of new vectors is rigorously controlled and has to be carried out in specially designed biosafety facilities.

The HIV virus that causes AIDS is a promising vector because it can infect nondividing cells, but it raises the specter of unwanted viral infection of the immune cells. However, safe HIV-based vectors have been genetically engineered so that they cannot grow when introduced into a patient. To maximize the rates of infectivity and integration, certain HIV genes have been mixed with those of adenoviruses to make a chimeric vector. At first glance this sounds even scarier. If a common cold virus could cause AIDS, the world AIDS epidemic would explode and endanger everybody. However, the scientists involved in this work are responsible, knowledgeable people who are well aware of potential dangers. They move cautiously with many layers of safeguards. Their experiments are all carefully scrutinized by panels of experts and involve much red tape.

The promise of gene therapy to treat Huntington's disease, cancer, and many other prevalent diseases for which little can be done is so great that it would be foolish not to pursue these avenues. It may be a long time before gene therapy is in widespread clinical application, but all radically new medical treatments take a while to be perfected. Many years passed before the dangers were recognized and avoided in such procedures as anesthetics, chemotherapy, and heart surgery. Physician scientists are still optimistic that gene therapy will one day be available for the critically ill.

Ethical panels have also been concerned that gene therapy might lead to irreversible hereditary changes. In other words, humans in the future might be affected by the medical treatment of their ancestors. Since most of the procedures aim at repairing genetic defects, this does not sound like such a bad thing. But if the techniques become so well established that minor defects are corrected by gene therapy, isn't there a danger that we will start to direct the course of evolution? It is not such a big step to go from treatment for hereditary dwarfism to treatment to increase the height of those who want to play professional basketball, or to go from treatment of mental disabilities to treatment for meekness. If these traits are passed on to offspring, then we might be in danger of having a population of aggressive giants.

For a transformed gene to enter the hereditary pool, it has to be incorporated into the germ line, the cells that give rise to eggs or sperm. Since most viruses stably integrate into replicating DNA, there is little danger that a woman given gene therapy would have eggs with the transgene, since most eggs do not divide after birth. But men continuously produce sperm, and if a transgene entered into a man's stem cell that differentiated into sperm, there is the possibility that one or more of his children would carry it as well. Sperm stem cells are held in the testes well protected from most agents in the blood. But the possibility exists. One solution would be to require all men who are given gene therapy to undergo vasectomy so as to preclude fathering children in the future. Since most patients now considered for clinical trials for gene therapy are extremely ill, such a decision would not be difficult. However, in the future, if gene therapy becomes more routine, the possibility of germ line transformation has to be taken very seriously. It could affect generations to come.

Similar concerns have surrounded the production of genetically modified plants. Techniques similar to those developed to introduce genes into patients have been used to transform crop strains with beneficial genes. The most successful application of these techniques has been the introduction of genes from a specific bacterium into corn, soy, cotton, and other major crop plants. The bacterial genes encode proteins that kill insects and worms but do not affect animals at all. These so-called Bt proteins form crystals that disrupt the digestive cells of invertebrates but are harmless to vertebrates. Crops that make their own Bt crystals are naturally resistant and do not have to be sprayed with strong chemical pesticides. The yield is improved and the environment protected from harsh chemical treatment. The financial benefits to the farmer clearly outweigh the cost of genetically modified seeds, and their use is increasing worldwide. Concerns have arisen that genetically modified plants will spread and take over much of the available habitat. There are worries that the genes might be passed to weeds as well. Insects and worms would then be strongly selected for resistance to Bt toxin, and after a while, there would be little advantage to having the Bt genes in the first place. All pests would be Bt resistant. At present, farmers are required to intersperse genetically modified crops with natural crops to stop their spread.

The ethical problems are not with the plants themselves. Who cares if an experimental seed fails? The company that develops them loses money, but there is always next year. The plants that are being genetically modified are themselves the product of thousands of years of controlled breeding by farmers to improve yield and tolerate harsh conditions. For a long time their evolution has been subject to unnatural selection by farmers to improve their crop yields. The problems are more political and financial than evolutionary. Nevertheless, various countries, especially those in Europe, have banned the importation of genetically modified foods and the use of genetically modified seeds. The fear of "Frankenfoods" is based on a misunderstanding of the genetic modifications themselves. The Bt genes cannot be passed on to those who eat them any more than any of the genes in food. However, it seems that a considerable number of European schoolchildren think that foods not genetically modified don't have genes!

Yellow journalism has fueled the fears of the man in the street that anything new is dangerous. Yet there was ready acceptance of potatoes and tomatoes from the New World five hundred years ago, and New Zealand kiwis and other exotic fruits found a ready market in Europe and elsewhere in more recent times. Crosses between tangerines and oranges sell well, as do seedless grapes. These are all genetically modified, artificial fruits, but this does not seem to bother anyone. It is just the foods from the people in white lab coats that are suspect. This does not seem to be a rational response. However, people opposed to genetically modified foods feel we are somehow interfering in natural processes. Many have a deep uneasiness about letting people take things in their own hands, especially things that affect heredity and may be long lasting. Although crops and domesticated animals have been modified for generations, the current, rapid pace of change is unsettling.

DOLLY THE SHEEP

For years, attempts to generate mice from enucleated eggs carrying somatic nuclei failed. The problem was thought to result from relatively stable modifications of DNA during differentiation of the donor somatic cells. Specific DNA sites near genes that are not needed in a given tissue are often modified, and this might result in faulty expression when the nucleus is required to generate all the diverse tissues of an embryo. Many scientists thought that these DNA modifications in somatic cells, referred to as genomic imprinting, might preclude cloning in mammals. However, in 1979 Karl Illmensee, a professor at the University of Geneva, reported that he had generated mice from eggs with the nuclei of four-day-old embryos. These startling results were soon brought into question and finally declared "scientifically worthless." All attempts to repeat the experiments failed for fifteen years.

On July 5, 1996, five years after Salva Luria died, the first mammal cloned from an adult cell was born. Dolly, as the little lamb was called, was the product of an enucleated egg from a Scottish blackface sheep that had received the nucleus from a cell taken from the udder of a six-year-old Finn Dorset ewe. Like other Finn Dorset sheep, Dolly was born pure white and stayed that way until she died, in 2003. The team led by Ian Wilmut at the Roslin Institute in Scotland used electric pulses to fuse somatic cells to enucleated eggs and transferred those that divided to virgin Finn Dorset surrogate mothers. In all, 276 eggs were used before the first successful clone, Dolly, was born (Wilmut et al. 1997). The technique has since been improved, but it still requires an inordinate number of eggs to generate a single cloned individual.

Dolly was shown to carry the exact same complement of nuclear genes as the sheep that provided the nucleus. She was an exact clone of the donor Finn Dorset sheep and could be considered her six-year-younger identical twin. The only genetic differences between the two were minor variations among the forty-two genes encoded by the mitochondrial genome. Mitochondria are maternally inherited, and so Dolly's came from the egg of the Scottish blackface ewe, while those of her "twin" were Finn Dorset.

Dolly grew up to be a healthy sheep and was bred on two occasions with a Welsh Mountain ram named David. She gave birth to Bonnie on April 13, 1998, and three more healthy lambs in 1999. She was clearly fertile. At the age of five, she showed signs of premature arthritis and died a few years later of progressive lung disease. An autopsy showed that she had pulmonary adenomatosis, a rare disease in sheep, but not so rare as to strongly indicate that it resulted from her unusual conception. Although sheep have a life expectancy of about twelve years, she died at the age of six and a half. It will be important to keep track of life expectancy in cloned animals.

The group at Roslin followed up with an even more dramatic feat (Schnieke et al. 1997). They grew skin cells from a sheep fetus in a Petri dish and transformed them with a human gene-encoding protein, factor IX. This is a commercially valuable protein that is given to hemophiliacs to promote blood clotting. If they could make sheep that had human factor IX in its milk, they would have a ready supply of inexpensive material for sale. Skin cells were selected in the lab for ones that had integrated the gene. These transformed cells were fused with enucleated eggs and implanted in surrogate mothers. One of them resulted in the birth of Polly in 1997. She was living proof that cells which had been transformed with foreign DNA could give rise to a viable animal. The commercial implications did not escape the attention of the pharmaceutical industry.

For a short while, Dolly was the only cloned mammal, but in the following year the group working with Ryuzo Yanagimachi in Hawaii reported that they could generate dozens of cloned mice by injecting nuclei from differentiated cells into enucleated eggs (Wakayama et al. 1998). The nuclei came from cells of a strain of mice with an agouti coat color, the eggs came from a black strain, and the foster mother came from a white strain. The pups were all agouti. Their DNA matched that of the donor, proving without doubt that they were generated by the cloning of an adult somatic cell. Since then the Yanagimachi lab has produced cloned mice at will.

The gates were open to cloning almost any mammal. Genetic Savings and Clone was set up to "resurrect" dead pets as a commercial venture. Their first success was CopyCat, a kitten born December 22, 2001, at College Station, Texas, with the genes of a calico cat named Rainbow. CopyCat, or CC as she is called, is also a calico, but her pattern of splotches and stripes is different from that of Rainbow. It has long been known that coat patterns of calico cats are determined by a random genetic inactivation process that happens in females. The owner of Rainbow should have known better, but she refused to take CC because she did not look exactly like her old cat. CC was taken in by Dewey Kraemer, one of the cloners.

The team at Texas A&M that cloned CC had previously cloned cattle, pigs, and goats. A cat was just another challenge. The team leaders, Mark Westhusin and Dewey Kraemer, are respected veterinarians interested in improving breeds and exploring cloning techniques. It took eighty-seven cloning attempts before CC was born, and the level of success will have to be improved before pet owners, other than the very rich, can afford to have their pets cloned, no matter how wonderful they may be. Westhusin and Kraemer do not feel that cloning domestic animals or pets is interfering with nature in any way, and they are now working to clone horses, dogs, and monkeys. However, like most people, they think that the idea of cloning a person is abhorrent.

HUMAN REPRODUCTIVE CLONING

The difference between therapeutic cloning, discussed in chapter 2, and reproductive cloning depends on what happens to the blastocysts. In therapeutic cloning they are used as a source of ES cells, while in reproductive cloning they are used to generate new individuals. In many countries, reproductive cloning is acceptable for farm animals and pets but is outlawed for humans. Scientific societies throughout the world have strongly endorsed a ban on human reproductive cloning until the consequences are better understood and people have had a chance to think about it.

California Senator Dianne Feinstein said in a prepared statement to the U.S. Senate on February 5, 2003, "Let's be very clear: human reproductive cloning is immoral and unethical. It must not be allowed under any circumstances" (http://feinstein .senate.gov/03Releases/r-cloning5.htm). Legislation to ban reproductive cloning in the United States became tangled up with attempts to ban therapeutic cloning partly due to confusion over the use of the word "cloning" in the very different procedures. The law is still not clear. Yet there is strong societal pressure to stop anyone from generating clones for the purpose of harvesting their tissues and organs whenever the donor might need them as transplants.

On the other hand, there are those who have argued that the cloning of an adult who is infertile should be permitted so that everyone can participate in the joy of parenting. "Everyone has the right to transmit their particular characteristics to their progeny, or to use cloning to reduce infertility," according to Dr. Severino Antinori. He is the director of the infertility unit at a center in Rome. In 1994 he gained attention when he implanted a donor's fertilized egg into a sixty-three-year-old woman, Rosana Della Cortes, making her the oldest known woman to give birth. He argued, "Generally, people are against human cloning, and I blame the media for pre-judging it. I want to bring society with me and persuade people that it is right in rare cases to help infertile couples." In 2002 he claimed he had a patient who was eight weeks pregnant with a cloned fetus, but there was no news of a birth. Several other quacks testified that they had produced human clones without any supporting scientific evidence.

Somewhere, sometime, human clones will be born. There does not seem to be any insurmountable obstacle to human cloning. Considering the low rate of success with cloning mice and farm animals, it may take a large number of women willing to act as surrogate mothers, but this could probably be arranged. The scientific societies that recommended a ban on human reproductive cloning not only voiced ethical concerns but also warned of the possibility of severe birth defects in individuals generated from enucleated eggs following somatic cell nuclear transfer. Experience with a variety of mammals has shown that most clones do not come to term and that birth defects are prevalent. It would be foolishly dangerous to jump to humans at this stage, before all the facts are known concerning cloning.

Studies at many levels are now in progress to improve the efficiency of cloning in test animals as well as to determine the genetic consequences in the offspring. Fairly soon it may be relatively safe to clone humans for reproductive purposes. The idea of generating a clone has been the subject of horror movies for some time. In one, it was assumed that cloned little Hitlers would all behave like the Nazi leader when they grew up. However, the Third Reich is history, and they might have become good shopkeepers or painters. The degree to which behavior is determined by nature or nurture has been debated since the beginning of written works. Does Fate control our lives or do we? Certainly genetics has something to do with it, but it is equally clear that life experiences make an enormous difference.

Identical twins come from the same fertilized egg and have identical genes. About four in every thousand births are identical twins that look alike but soon start to take on individual traits.

Even at birth their fingerprints are different. The direction of the hair whorl on the crown of the head can be clockwise in one and counterclockwise in the other. One may be right-handed and the other left-handed. Certain processes in embryogenesis have a degree of randomness in them that can affect the individual. Remember that CC the kitten did not have the coat pattern of the calico cat that provided the somatic nucleus. Both were female, but the pattern of X-inactivation in CC was different from that in Rainbow. Twins who grow up together usually have a special bond with each other that can last throughout life. They laugh at the same jokes and pick the same ties. Those raised separately can end up with quite different personalities, although they always look surprisingly similar. It is not all genetics.

A "twin" generated by human reproductive cloning from the nucleus of an adult obviously does not have the same life experiences as the donor. They are at least a generation apart and have different mothers. As a cloned boy grows up, he will notice that he not only looks like a younger version of his "father" but also shares certain characteristics, such as intelligence, aggressiveness, or even a predilection for chocolate ice cream. When he learns of his origin, he may realize that he can predict his life course better than others. If his "father" had a heart attack at the age of fifty, the younger man may worry that he too will have a heart attack in middle age. In fact, his chances of such a problem are greater than those of the son of a couple who conceived him naturally, even if the biological father also suffered a heart attack at fifty. Knowing what is in store can be useful or disturbing. The cloned individual may want to have frequent medical checkups during middle age and lead a healthy life that reduces the chances of a heart attack. Or he may make choices in early life on the basis of knowing he may die young. They might not all be good choices.

GENETIC MODIFICATION OF *HOMO SAPIENS*

Polly the Sheep was generated from skin cells after the introduction of the gene coding for human factor IX. If it could be done for a sheep, there is no reason to think it could not be done for a human. And this raises all the red flags that worried me more than forty years ago in discussions with Salva Luria. It seems that it might be possible to direct human evolution in my lifetime, or at least that of my grandchildren. And we still have not had a broad public discussion of the advantages and disadvantages.

Doctors who are concerned with devastating diseases that result from specific faulty genes usually see no harm in replacing them with a good copy if it would help their patient. They would happily generate blastocysts from enucleated eggs carrying nuclei from some of the patient's cells. ES cells could then be established and transformed with a good copy of the pertinent gene. Returning these cells to the patient might cure the disease. Somewhat later, the patient might want a healthy child and convince his or her doctor to thaw one of the blastocysts made by somatic cell nuclear transfer and allow it to come to term. If the patient is a woman, she could carry it herself. She could then have a child that did not suffer from the genetic defect that she carried. The child would be a clone, but at least it would be healthy clone. When the child grew up, it could pass on the healthy gene to its own children. This all sounds fairly innocent, but it could have far-reaching effects if germ line therapy became prevalent.

The societal problems associated with genetically modified clones could be avoided by using blastocysts generated from in vitro fertilization of an egg with a sperm. Cells of the resulting blastocysts would have the same mix of parental genes as those of any child and would not be direct copies of either the father or the mother. Cells could be recovered from the blastocysts and grown in Petri dishes before being transformed with genes to replace faulty ones or to provide any desired trait known to derive from a single gene. One of the genetically improved cells could then be fused with an enucleated egg and returned to the mother. Nine months later she might give birth, gratefully, to a child free of the genetic defect she knew she was carrying.

Eradicating a deleterious gene sounds like a good policy. It is a lot of work to replace a single bad gene, but the end result is that the problem would be avoided for generations. However, genetic techniques are not yet well enough understood to suggest that efforts should be made to improve the gene pool of the species. The situation may change during this century, and discussions concerning the consequences of such genetic modification should start before the tools are in place. Almost every scenario that has been considered is fraught with danger.

Some cases are so dire, it is worth considering whether genetic modification might be warranted. What if the AIDS pandemic that is killing millions every year got worse and threatened the survival of the species? If the only way to stop it were to resort to genetic modification, should we proceed? The HIV virus that causes AIDS can evade almost any treatment because of the high mutation rate of its RNA genome. No drug stops HIV in its tracks. Some drug cocktails slow it down, but it almost invariably kills. Drug-resistant variants arise over time and overwhelm the patient. Vaccines have never been found to work, and if they did, it would not be long until an HIV mutant would be selected that was resistant. It is a serious problem. If I were convinced that HIV or other such viruses were likely to lead to the extinction of the human species or even generate a serious bottleneck, I would advocate immediate germ line intervention. Although it might not be possible to predict all the consequences, generating a group of resistant individuals who could survive and multiply might save the species. In my opinion, extinction should be avoided by all means.

Could HIV be stopped? Techniques for silencing genes with an antisense copy are well established in model systems. If a copy of the gene is inverted relative to the signals that indicate the direction of transcription, then the product is the same as the original gene, except it is backward. Since the two strands of DNA are complementary, the antisense RNA will be complementary to the normal messenger RNA. Just as complementary DNA strands recognize each other and zip up into a double-stranded molecule, complementary RNA strands will form double-stranded RNA. Cells cannot translate double-stranded RNA into protein, and the targeted gene is silenced. When applied to HIV, any of the nine genes could be chosen to generate antisense copies. It might be advantageous to string several genes together and make a tandem-array antisense to completely prevent HIV from making critical proteins and growing.

The genetic engineering could be done with well-established techniques, and the resulting constructs could be transformed into ES cells from in vitro fertilized eggs. They would have the normal diploid content of genes as well as the antisense construct in their chromosomes. When the transformed cells are fused with an enucleated egg, they could give rise to an embryo. The child would be born immune to HIV and could later pass on the resistance to offspring. There would be no way the virus could mutate to avoid being silenced by the antisense RNA, since double strands will form even when there are mismatches.

In the worst case scenario, every one might die except those who carried the HIV-antisense gene. The population would go through a severe bottleneck and lose most of the genetic diversity that has accumulated over the years. Inbreeding can produce a high proportion of mutants, and even those who seem all right might turn out to be especially susceptible to a lethal disease. Animal species that have gone through near-extinction, such as the cheetah, have dramatically reduced genetic diversity and are at risk of going extinct any time when the environment changes. To keep genetic diversity of the human species as high as possible, couples from Asian, European, African, Native American, Polynesian, and other cultures could be recruited to have HIV-resistant children. If there were enough time, millions might be made HIV resistant. Further reproduction could spread the anti-sense gene widely before the pandemic took its toll.

Even if we do not face extinction, should we consider using molecular genetics for minor improvements in the species? Many people have commented on how skin color divides populations, pitting one group against another. Skin color is a noticeable but trivial genetic difference in the human species and might be dispensed with to increase harmony. People often show solidarity with those who look most similar to themselves, and if we all looked more alike, we might get along better. Almost everyone knows that we are all the same under our skin and can conceive children together, but knowing we are all one species has not stopped wars and exploitation fueled by differences in appearance. It might help if this marked difference were reduced. It would not be hard to transform cells with genes for increased or decreased production of melanin, the dark pigment generated from the amino acid tyrosine by melanocytes in skin. The production of melanin is controlled by the enzyme tyrosinase. Research could be directed at the mechanisms that control proliferation of melanocytes or tyrosinase levels, so that the appropriate genetic construct could be introduced into the cells before they were fused with an enucleated egg.

People with very low tyrosinase have completely white skin and hair—they are albinos. An estimated eighteen thousand people with albinism now live in the Unites States. All of them have trouble with their eyesight, since the lack of pigment affects development of their retinas. Many cannot see well enough to drive a car. The problems faced by albinos in Africa are even greater. The contrast with the surrounding dark-skinned population often results in social ostracism. Moreover, they suffer terribly from sunburn in tropical countries. Almost any albino would be in favor of ensuring that everyone was born with genes that made enough pigment. But how much is enough?

Northern European populations have fairly little tyrosinase and produce fewer melanocytes as the result of selection for light skin. A mutation in the gene (SLC24A5) that encodes a membrane protein in melanosomes is prevalent in Europeans but essentially absent in Africans (Lamason et al. 2005). It appears that when the original dark-skinned humans migrated into the Caucasus, Siberia, and Scandinavia about ten thousand to twenty thousand years ago, they initially suffered from the bone-deforming disease rickets, which results from insufficient vitamin D. Sunlight is required to make vitamin D, and there isn't much sunlight during the winter months in the Far North. When one's vitamin D level is low, bones soften and can be deformed during childhood. Since dark skin blocks much of the ultraviolet light needed to make vitamin D, the original population of migrants must have included many who were crippled. However, there were a few individuals with lighter skin who carried the variant gene and so could use the weak northern sunlight to make enough vitamin D. Over time these healthy few multiplied and spread out over their territory. Later they spread back to the south and became the dominant population of Europe. Now there are "white" people all over the globe. There are no serious disadvantages to being fairly light skinned in New York, Rome, or

Kiev. In the Caribbean or Africa, a little sunscreen can protect "whites" from sunburn.

A committee entrusted with determining how the human genome should be modified to blend the range of skin colors might aim for a Mediterranean or Polynesian swarthiness. It would be sufficient to protect from sunburn, and yet light enough to cause no harm for those living in Sweden or Alaska. Would we all want to be the same color? An emphasis on skin color is deeply embedded in the historical European self-image, and some people would strongly resist any change. And yet the drawbacks to colonial racism are deeply divisive in the world. In both North and South America, the socioeconomic lines are often evident from the color of the population. A color-blind society would be much preferable and might be more easily achieved if the differences were not so readily noticeable.

Once the desired skin tone was chosen, the appropriately engineered gene could be transformed into somatic cells before fusing them with enucleated eggs. The resulting individuals would all have about the same level of pigmentation and would pass this along to their children. However, it is impossible to transform more than a tiny proportion of all the individuals conceived in a year. If the goal is to genetically modify *Homo sapiens* so that skin color becomes more homogenous, it would be necessary to link this trait with some highly desirable property that would help to spread the modified gene throughout the population. If this selection were sufficiently strong, then in a few dozen generations most people would carry the desired gene. What trait is so highly coveted that it would surely spread rapidly in the population, taking the skin color gene with it? Longevity.

Almost everyone would like to live longer. Even if genetic modification could not help them, they would still want their children and grandchildren to live to two hundred years of age. We are all suffering from genetically programmed aging, and few of us will live to be a hundred. Given the chance to live longer, we would jump for it. It is also likely that a woman of normal life span would chose to conceive children with a man of extended life span so as to give her children and grandchildren longer lives. The same goes for a man of normal life span looking for a mate with extended life span. It would not take long before the demographics showed that the population as a whole was living longer.

We don't know which genes control the human life span, but there is strong evidence in model organisms that changes in a few genes could extend life by 50 percent and sometimes even double it. SOD (super oxide dismutase) is an enzyme that protects cells from oxidative damage and might lead to longer life if engineered to function more efficiently. There is also genetic evidence that lowering the body temperature of mice by 0.5°C can prolong life up to 15 percent. That would mean 8 to 10 more years if it worked in humans. If we just stayed a little cooler, we might live longer. However, we want to feel like 50 when we are 150 or, better yet, feel like 20 when we are 200. So it will be important to carefully modulate the genes so that we not only live longer but also live well for longer. There is a lot to learn before aging is cured.

Even if the life span could be extended, it is not clear that this would be a good idea. It would only worsen the problem of world population, which is already beyond the carrying capacity of the planet as it now stands, at 6.6 billion. If people did not die until they were 200, the population would double again, especially if people continued to have children when they were over 100. The only solution would be to have rigorous population control coupled with longevity. How to modify *Homo sapiens* to regulate procreation in a manner that could adapt to unforeseen challenges has not even been fantasized. This is all beginning to sound like science fiction.

While we are a long way from being able to rationally modify *Homo sapiens*, the underlying modification techniques are improving all the time. It is no longer a foolish waste of time to consider what we might want the species to be in ten or twenty generations. It may take hundreds of years to reach a broad consensus. Would future generations be grateful that they had received genes for increased radiation resistance if they lived in a world polluted with the waste of nuclear power plants? Would they think it better that they all were similar in appearance and lived for 200 years?

Who is going to judge? Such matters should not be left up to scientists, most of whom are interested only in seeing how far they can take genetic engineering. Nor should it be decided by popular vote when the citizens do not understand the underlying population genetics, molecular biology, and medical facts. Genetic enhancements designed to give a short-term competitive edge to a small group of people might have long-term effects on the species. Only in the last 25 years have people come to understand that their hereditary traits are controlled by the sequence of their DNA. Even when the subtle interplay of genes and proteins is common knowledge, the wisdom to decide on the fate of humankind will depend on a deeper and more rational appreciation of morality and ethical values. For the time being, it is best to refrain from tinkering with our species, which has managed to survive this far in evolution. In my opinion, a ban on germ line transformation should be strictly enforced throughout the world until the day comes when germ line transformation is considered essential or a consensus on its value is reached.

SYNTHETIC GENOMES

Like a witches' cauldron, new techniques are brewing that may lead to synthetic genomes. Synthesizing genomes from scratch was inconceivable a few years ago, but recent advances in the chemical synthesis of exact sequences of fifty, one hundred, or even two hundred DNA bases in a row, which can be put together to form longer strands, has made such a goal achievable. In 2002 a group led by Eckard Wimmer at Stony Brook, New York, synthesized the 7,411 base pair (bp) genome of polio virus and showed it was infectious (Cello, Paul, and Wimmer 2002). They got the sequence off the Web and just built fragments that were 70 base pairs long. They put 110 of them together in the proper order and made the virus. Just when polio is on the brink of being eradicated, it is frightening to think that a terrorist or bio-hacker might decide to brew up a batch. Wimmer said, "The world had better be prepared."

Other more deadly viruses are within reach. The genomes of Ebola and smallpox are both about 19,000 bp and so could be synthesized using present techniques, although it would be a difficult and laborious task. People infected with Ebola die within a few days as a result of fluids leaking through their epithelial membranes. Smallpox is highly contagious and often kills those who had not been previously immunized. Routine immunization in the United States ended in 1972 after smallpox had been eradicated in the country. Both Ebola and smallpox are on the list of bioterror agents, and all stocks are kept in carefully guarded laboratories. But they could be synthesized in clandestine laboratories and let loose.

Viruses are not alive, and so the synthesis of polio virus is not the same as the synthesis of life. Viruses are inert until they infect a cell, where they find the machinery for transcription and translation of their genes. They are like blueprints without a construction crew. Each virus has its own set of instructions, which may be benign or deadly to the host cell. Even innocuous viruses could be tampered with to give them new properties. George Church, a professor of biology at Harvard, chaired a panel on the problems and opportunities of DNA synthesis at the "Synthetic Biology Conference" held at the Massachusetts Institute of Technology in 2004. He said, "We will still create things that do not have the properties that we thought they would." Tampering with viruses might be a dangerous adventure.

Others are not so pessimistic. J. Craig Venter, who led the private effort to sequence the human genome, has said, "The field of synthetic genomics has the potential for groundbreaking scientific advances, including the development of alternative energy sources and the production of new vaccines and pharmaceuticals." Technological innovation is increasing so fast that, in a relatively short time, we may learn the dangers of genetic manipulation so that we can avoid them. The convergence of information science, computing, and biotechnology may lead to a significant increase in the pace of technological expansion. There are people who really love the idea of synthetic biology, because they see it as the next stepping-stone toward bionic everything.

A gift from the Bill and Melinda Gates Foundation helped establish a Department of Synthetic Biology at the University of California, Berkeley in 2004. The department is exploring ways of biologically synthesizing anticancer drugs, new fuels, and compounds for the electronics industry. Plans are also in the works to modify metabolic pathways for specific uses. Other laboratories have already engineered *Escherichia coli (E. coli)* cells to display spontaneous oscillations and respond to signals in the environment by switching to a new state. Circuits that pass a signal except when two signals are simultaneously presented (a

NAND gate, in computer-speak), and ones that pass a signal except when one or the other of two inputs is strong (a NOR gate, in computer-speak), have been made that work fairly well in bacteria and yeasts. Likewise, circuits that pass a signal only when two inputs are simultaneously strong (an AND gate) have been constructed. Attempts to make similar devices in animal cells are under way. These are not yet synthetic life, but they are highly engineered life-forms. Some synthetic biologists want to build systems from the bottom up so they completely understand the workings. They want to explore the properties of living things that work as an engineer would expect because they were built by an engineer.

Further technological innovations are being developed that may soon lead to the synthesis of complete bacterial genomes or even whole chromosomes. Since the problems of synthesis get bigger as the size of the genome gets bigger, synthesizing the 4,639,221 base pairs (bp) of DNA in the *E. coli* genome will take some time. Meanwhile, Fred Blattner at the University of Wisconsin is in the process of removing pieces of the *E. coli* genome to see how small he can get it and still have the bug grow in the laboratory (Posfai et al. 2006). So far he has cut out about 700,000 bp and finds that it grows well with only 3,500 genes. He plans on continuing to pare down the genome until the minimal essential genome is reached. This will make a good starting point for adding designer genes to carry out specific functions.

Ham Smith, a Nobel Prize winner at the J. Craig Venter Institute, started with the much smaller genome of *Mycoplasma genitalium*, a bacterium that lives inside of larger eukaryotic cells. He was part of the team at The Institute for Genomic Research (TIGR) that sequenced the 578,000 bp of the *M. genitalium* genome in a two-week period in 1995 to demonstrate the power of whole genome sequencing (Fraser et al. 1995). They found that the bacteria carried only 470 genes, and they have been able to remove 100 of these genes without killing the organism. When they have it down to the bare minimum, they aim to synthesize and assemble the genes into a viable genome. This is the next logical step in genome biology, as it is the only way to better understand the minimal components consistent with cellular life. One of the tenets of chemistry states that to prove true understanding of a structure one must be able to synthesize it. The same can be said for genomes. Ham Smith has indicated that, to avoid possible runaway monsters, they will synthesize only organisms that completely lack the ability to survive outside the lab.

The plan is to insert the synthetic genome into a bacterial cell that has lost its own DNA and show that the bug is now able to grow. Even if they are successful, they will have synthesized not life but only a genome. It will be no more alive than a virus, since it will require all the transcription and translation machinery of the cell. However, once a genome has been synthesized, it is fairly easy to make modifications to test hypotheses and give the bacteria novel properties. The technology is changing so rapidly that it is impossible to say where the limits might be in twenty to fifty years. Will it be possible to synthesize the 342,000,000 bp genome of a puffer fish?

It would be exciting if a fish could be made with a synthetic genome, but it would not be as if life had been synthesized. Only the genome would be made in the lab, while the egg would have to come from a real live fish. The machinery present in the enucleated egg would be essential for the readout of the genetic instruction. If both male and female fish were made, then the progeny of their eggs and sperm would be wholly dictated by the sequence of the synthesized genomes.

The human genome is ten times bigger than the genome of a puffer fish and presents a series of presently insurmountable problems. Which is all to the good, because we have not yet decided as a society if it should even be contemplated.

Tampering with our genes would affect generations to come, and we don't even know what is in store for our grandchildren. We might provide them with needed protection, or we might cause terrible trouble. Considering the level of complexity in the circuit that regulates the diet of the simple bacterium *E. coli*, discussed at the beginning of this chapter, there is a good chance that changing human genes would have unforeseen effects. The networks connecting genes and proteins are so poorly understood that we cannot predict with any certainty that replacing a faulty gene or overexpressing a protective gene would not have undesirable consequences. Errors would be hard to correct and might not be corrected in time. Even a single small change could lead us into the treacherous terrain of more substantive changes. Considering the arrogant ignorance often displayed by ruling groups, the species might suffer irreversible damage. We should be very wary of starting out on such an adventure. I know of no one who would take the responsibility for the fate of humankind on his or her shoulders. If we are going to open Pandora's box, let's do it very carefully.

Even without synthesizing designer genomes, knowing the genome sequence inherited by each of us raises all sorts of problems (Kitcher 2001). Widespread prenatal sequencing for disease genes has enormously increased the number of instances in which it can be predicted with confidence that the outcome of a given pregnancy will be tragic. There is usually nothing that can be done to fix the problem, and if the embryo is brought to term, the infant will be severely disabled and the parents devastated. Terminating the pregnancy by aborting the fetus is often considered the most humane response. However, many people feel that the life of a fetus is as important as the life of its mother and would outlaw all abortions. There is a slippery slope from god-awful genes to simply unwanted traits that has to be negotiated. Moreover, some genes, such as the one for right- or left-handedness, do not dictate the outcome but leave it up to chance. Genes affecting personality and social behavior may predispose an individual to turn out one way or the other but without imposing strict limits. These are the subjects of the following chapter. They can be fully considered only in the context of understanding the biological basis of consciousness, memory, quality of life, and cooperativity, which are discussed in later chapters.

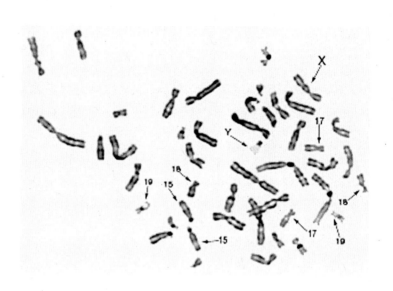

FOUR

GENOMIC INFORMATION

GENETIC MAPS TELL YOU WHERE THE GENES ARE ALONG THE CHRO-
mosomes and organize all the available information in a manner
that can be readily seen. I love maps. A few years ago I was at El
Escorial palace near Madrid in Spain, where in the sixteenth century King Felipe II covered the walls of a map room with colorful depictions of the new lands. The explorers who sailed the
world in the fifteenth and sixteenth centuries came home with
new information that each year led to improved charts and
global maps. Europeans learned of whole new continents as well
as mountains, rivers, and islands. These maps have errors of
placement for some islands, and many names have changed, but
you can still recognize the world. I would have enjoyed sailing

FIGURE 4.0 Human chromosomes. Hereditary information is encoded in
the sequence of DNA found in twenty-two pairs of autosomes and a pair of
sex chromosomes. Chromosomes 15, 17, 18, and 19 and the X and Y sex
chromosomes are indicated. Courtesy of the Lawrence Berkeley National
Laboratory Technology Transfer.

with Magellan in 1519. They sailed west and returned from the east. They charted the tip of South America, the Philippines, Borneo, and many things in between. Now the land everywhere in the world is precisely mapped, with a margin of error of no more than a few meters, using images sent down from satellites. Navigation may be much easier, but the fun is over.

Genomes are the terra incognita of this century. I remember when the first complete map of the bacterium *E. coli* was published in 1964. It was based on recombinational frequencies between gene markers and had many gaps and errors, but it put genes in their places. Over the years it was improved and refined using a variety of techniques (Taylor and Trotter 1967). Then in 1997 a team led by Fred Blattner sequenced the complete *E. coli* genome. The order of the 4,639,222 bases in DNA was determined and mapping was finished (Blattner et al. 1997). During this time I was working with the soil amoeba *Dictyostelium discoideum* and wanted to generate a genetic map for each of its six chromosomes. Initially, genetic markers were assigned to each chromosome by following their segregation from diploid strains. By 1990 it was possible to map genes using specific sequences.

We built up maps of each of the six chromosomes with markers placed every hundred genes or so. The maps were subsequently refined by using a technique that Paul Dear invented, called HAPPY mapping, which relies on physically breaking DNA and seeing which genes are still found together on the small pieces. These maps had markers placed about every five genes. Since their order and spacing was known, they could be used to assemble the DNA sequence, which necessarily came in pieces covering only a dozen or so genes. The sequence of the 34 million bases of the *Dictyostelium* genome was completed with high accuracy and reliability in 2005 (Eichinger et al. 2005). Mapping was finished; it had been brought to the ultimate level of the single base. The next challenge was to figure out what it all meant.

Sequencing of the human genome followed a similar path. In 2003, working drafts of the human genome sequence were presented simultaneously by the federally funded laboratories and the private company Celera, which was headed by Craig Venter. The two maps agreed on most points, but there were significant differences. Over the next few years, each chromosome was subjected to intense scrutiny by the publicly funded labs, and the "final" sequences were deposited in easily accessible files. The trouble with huge projects like the human genome sequence is that minor points are set aside until another day. For instance, the sequence was generated from the DNA of multiple donors, both male and female, and each donor gave two copies of each gene, the maternal and the paternal copies. There are many slight differences in the copies of different individuals that were covered over by the most common sequences. So it really isn't the sequence of any one person; it is the average sequence of the donors. Efforts are now being made to sequence specific regions from identified individuals to see just how much variability there is and to provide markers along each chromosome. The full sequencing of individual genomes will probably not be routine before novel sequencing techniques are developed. There are promising techniques, but they are not quite ready for prime time.

INDIVIDUALITY

If a person were handed the sequence of his genome, what would he see? The huge line of As, Ts, Gs, and Cs would stretch into the distance for 3 billion bases. There would be no feeling of self, although this sequence is what defines an individual. Moreover, the sequence would have to be stored and displayed in a computer to be of any use, and it is hard to relate to a computer file. A printed version of a personal Book of Life would require over thirty thousand pages of small type to cover the twenty-three chromosomes. A computer would have to be used to compare the sequence to the generic human genome sequence and flag minor variations. It is likely that all twenty-five thousand genes would be found in their expected places. In between genes, there would be long regions of apparently meaningless sequences. Some of the bases in these wastelands might be important for one reason or another, but at the moment we don't know where to look for them.

One use for a personalized sequence is tracing family history. Genealogy fascinates many people because it gives them a sense of belonging to an extended family. Rare differences in sequence arise as errors in replication of the genome, and they are carried forward through the generations, so that you can see traces of your great-great-grandparents. It is possible to recognize the Polish variation, the Irish variation, the Congolese variation, the Thai variation, and so on, and know where your ancestors came from. Most people are surprised when they see a part of their own sequence and find out about their genetic mix—they never knew they had a Greek or Chinese ancestor, but the telltale variations are there. The evidence would be even stronger if the complete sequence could be inspected.

Among the twenty-five thousand genes, there might be a few unusual DNA sequences that would result in amino acid changes in proteins. It has been estimated that most people harbor variants in about 1 percent of their proteins. Many variants have no consequences at all, even if both copies of the gene are mutant. Among the others, most variants are recessive and have no consequences as long as there is one good copy of the gene. Only when both copies harbor a deleterious mutation is the individual affected. The root cause of most diseases are still unknown, so the person may have inherited a bad gene but will know it only when he or she falls ill.

Without further analysis the differences in DNA themselves would not mean that the individual was at risk for some disease or malfunction. Only if one of the thousand or so mutations known to cause hereditary diseases was found in the genome would there be reason for concern. Even if we knew how to read the Book of Life and catch all the nuances, it would tell us only whether our cells and our bodies were put together well. It would not tell us if we would be hungry or sad, how our choices would turn out, or whether our next grant application would be funded—the important things in our daily lives. It *can* tell us if we were dealt a bad hand in the genetic card game and carry a mutation likely to cause us misery. In most cases, there won't be much we can do about it.

For patients with cystic fibrosis to learn that they have one of the twenty-four common mutations in the CFTR gene is not very comforting. They knew they were sick before the tests, and there is nothing that can be done to fix the mutation. The CFTR gene makes a transporter that controls the entry of chloride ions into lung and pancreatic cells. Do patients really care that a phenylalanine is missing at position 508? Their congested lungs, coughing, wheezing, and salty sweat would be indication enough. They know they will die young, and that nothing can be done about it. Couples with any family history of cystic fibrosis are encouraged to have their CFTR genes sequenced—just the gene, not the whole genome. If both parents carry one bad copy of CFTR and one good copy, they will be healthy but their children will have a one in four chance of developing the disease. When they have been informed of the medical problems of cystic fibrosis, they may decide not to have children, or they may go ahead and conceive but have prenatal genetic testing. The American College of Obstetricians and Gynecologists now recommends that all couples who are planning pregnancy be screened for cystic fibrosis. Although the chances of being a carrier are low, the consequences of having a child with cystic fibrosis are so devastating that it is best to catch it before an affected child is born.

Some populations carry an unusually high proportion of defective genes. Again, a defective gene causes no problems for people who also have a normal copy of it. But when a child inherits the bad copy from each parent, it can be tragic. Genes leading to several well-known hereditary diseases, such as Tay-Sachs disease and Gaucher disease, are especially prevalent among Ashkenazi Jews. Those who inherit the mutated form of the Tay-Sachs gene encoding b-hexosaminidase from one parent and the normal gene from the other parent are healthy and even have an advantage because they are much less susceptible to tuberculosis. Those who inherit the mutated form from both parents accumulate excess fatty acids in their nerve cells and start to lose brain functions within six months of birth. They seldom live more than a few years. It has been estimated that 10 percent of Ashkenazi Jews carry one or more genes that can lead to birth defects and early death. Members of this population are encouraged to be tested for these genes before they plan on having children. If both are found to be carriers of a gene for a devastating disease, they should check the genetic makeup of the embryo in the first trimester. If the fetus is found to carry two copies of the defective version of the gene, the most rational response is to abort the pregnancy to avoid untold suffering.

Another of these horrible diseases, Canavan disease, is caused by deficiency of the enzyme aspartoacylase (ASPA). A deficiency in ASPA results in the buildup of a chemical that destroys the "white matter" in the brain. Slowly the brain deteriorates and the child dies. The gene for ASPA has been mapped to chromosome 17, and there is a straightforward test for the mutated form.

About 2 percent of European Jews carry the mutated ASPA gene. They are healthy because the other copy of the gene is normal and provides all the ASPA they need. If both copies harbor the recessive mutation, the individual will develop Canavan disease. Babies with this disease appear perfectly fine for the first few months, but then show signs of a lack of head control. Motor skills continue to decline, and the children go blind, although their hearing remains sharp. They exhibit feeding problems, stiffness, and seizures before finally dying. A friend of mine and his wife had a child with this disease, and I know how horrible it can be to watch the child waste away. They would have done anything to cure their child, but there was no cure. They had not been tested for ASPA because one of them was not of Jewish origin and they thought the chance was too low to worry about. Before conceiving another child, they were both tested for a wide range of potentially deleterious genes and were prepared to abort a fetus if it turned out to carry genes for Canavan or any other horrible hereditary disease. They have two healthy children.

The state of New York now routinely screens newborns for fifty different inherited diseases. However, it is too late to help most of the unfortunate infants found to carry genes causing serious disabilities. Their future can be foretold with certainty, but seldom can much be done to change it.

ABORTION

Prenatal testing of a fetus for genes leading to birth defects and serious hereditary diseases can be done only after fifteen to eighteen weeks of pregnancy. At that time a few cells can be isolated from the amniotic fluid surrounding the fetus and characterized for chromosomal abnormalities and specific gene defects. If the prognosis is dire, abortion may be recommended. If the analyses indicate that the child will have Down's syndrome, spina bifida, Canavan, or any other devastating hereditary defect, the parents are often advised to consider terminating the pregnancy rather than give birth to a child who can have only a short, highly constrained, and painful existence. Abortion procedures are relatively safe and simple when carried out in clinics. There is an emotional toll on the mother and father, but it is minor compared to that resulting from the loss of a teenager whom the parents had loved and nurtured for years.

However, feelings about abortion run high in many societies (Gazzaniga 2005). There are some who feel that any intervention is wrong, and that nature should be allowed to run its course, while others feel it is unfair to knowingly bring a defective child into the world. Much of the world suffers from a conflict between pro-life and pro-choice advocates. Pro-life advocates would like to ban all or most abortions, while abortion-on-demand advocates argue that it is a matter that should be decided only by the woman. Those who favor choice feel that it is not the government's business to interfere one way or the other in such a personal matter. They argue that outlawing abortion can punish the victims of rape and incest and might threaten the woman's health. Even if the harm is only to her mental health, they feel this is sufficient grounds for abortion. Even more forceful advocates of free choice point out that an unwanted child will be more likely to be poorly raised, neglected, and even abused. Many will grow up at the margins of society and lead antisocial lives, often ending up in jail. The pro-life advocates counter that life begins at conception and that abortion is akin to murder. They feel that all potential human life is sacred and must be protected. The question is, when does a human life begin?

For most of its history, the Catholic Church had no objection to abortion before "quickening"—that is, when, at about twenty weeks of gestation, the mother could feel the child move within her. Until recently, for the first few months a woman had little idea if she was pregnant or not. Only when she was surprised by small butterfly flutterings in her belly would she conclude that the recent missed periods might indicate that she was with child. If the pregnancy was unwanted, such as might be the case if her husband had been away for a few months, women resorted to herbal potions and drugs much like the morning-after pill RU486, which induces miscarriages. In 1869 the Catholic Church came out against all abortions at the request of Napoleon III of France, where the population was declining. Pope Pious IX declared that life begins at conception and had to be protected thereafter. In the Jewish tradition the fetus is part of the mother until thirty days after birth, and she can do as she wishes; however, there are passages in the scriptures that consider the first breath as the start of life. In either case, abortion is a private matter for the mother. Although there are many different readings, Hindu teachings appear to allow abortion until the fifth month, when movements of the fetus can be felt on the swelling belly.

In 1995 Pope John Paul II declared that the church's teaching on abortion "is unchanged and unchangeable. Therefore, by the authority which Christ conferred upon Peter and his successors . . . I declare that direct abortion, that is, abortion willed as an end or as a means, always constitutes a grave moral disorder, since it is the deliberate killing of an innocent human being. This doctrine is based upon the natural law and upon the written word of God, is transmitted by the Church's tradition and taught by the ordinary and universal magisterium. No circumstance, no purpose, no law whatsoever can ever make licit an act which is intrinsically illicit, since it is contrary to the law of God which is written in every human heart, knowable by reason itself, and proclaimed by the Church" (*Evangelium Vitae* 62). This seems pretty strong, but abortion has been legal in Italy since 1978 and was supported by 80 percent of the voters in a 1981 referendum. About 1 percent of the women of reproductive age in Italy have legal abortions each year. In both Sweden and the United States, the abortion rate is 2 percent among women aged fifteen to forty-four.

In the last thirty years, pregnancy kits and ultrasound scans have facilitated early detection. There is something about seeing a peanut-shaped mass in the womb that stirs parental feelings. When a woman knows she is pregnant, she often takes better care of her health to protect the fetus. The bond between mother and future child grows stronger with every visit to the clinic and every new sonogram showing the developing limbs and beating heart. But if the sonogram shows gross malformation of the fetus, shouldn't the mother have the right to terminate the pregnancy?

Most countries have come to accept abortion as a woman's right, and some cover the expenses when necessary. Starting in the 1970s the Greek Orthodox Church and pediatricians on Cyprus encouraged prenatal diagnosis for beta-thalassemia. At that time, 1 in every 158 infants born on the island was afflicted with this blood disease, which results from defects in hemoglobin. The afflicted are sickly, anemic children who usually die before reaching the age of five. The program of population screening and abortion encouraged by the Cypriot Church and the government resulted in almost complete prevention of new births of children with thalassemia in both the Greek and Turkish populations on the island.

Just to the north, on the mainland of Turkey, the Regional Health Administration of the Denizli carried out a pilot program in 1995 aimed at reducing the incidence of beta-thalassemia. All couples who applied for a marriage license were screened for the recessive trait. Couples at risk were counseled and offered prenatal diagnosis and abortion of an affected fetus. A total of 9,902 couples were screened, and the trait was found in 514 individuals. In 15 of the couples, both partners were carriers. Seven of these couples decided not to have children. Prenatal diagnosis later discovered an affected fetus among the other couples, and the pregnancy was terminated by elective abortion. At no time in this pilot study was there any coercion or pressure brought to bear on couples to abort an affected fetus; but when the medical facts were explained to the couples, they chose to abort or refrain from having children.

As the number of mutations linked to hereditary diseases increases in the coming years, prenatal testing will turn up a growing number of cases where the prognosis is poor. We will be able to test broadly, but only rarely will we be able to offer any help when a serious defect is found. The criterion for terminating a pregnancy may slip down the slope to the point where it is at the mercy of current fads and prejudices. The long shadow of eugenics falls across the field, reminding us of past misuses of genetic arguments.

Even now some fetuses are aborted only because they are female. If brought to term, they would be healthy girls, but the parents want a boy child. This seemingly irrational behavior results from the strong pressures for male offspring in some cultures. Abortion is so lightly considered in these cultures that, when some people find out that their fetus is female, the pregnancy is terminated. Will there come a time when a fetus is aborted just because it carries a few genes from the side of the family that was not very successful? Abortion to avoid giving birth to a seriously defective child or to protect the mother is one matter; abortion to fit the whims of the parents is another. However, it is difficult to regulate one without affecting the other.

Perhaps the choice has to be left up to the woman, but she should be educated to know the difference between medical facts and petty preferences.

If most genetic problems detected by prenatal testing are beyond our ability to alleviate, and the only humane thing to do is to terminate the pregnancy, perhaps we should not continue testing for so many genes. Without the ability to terminate the pregnancy, testing would not result in any decrease in the number of afflicted individuals. However, it can be argued that no matter what one's stance on abortion, it is usually better to know beforehand what problems to expect. For instance, there is a straightforward test for the rare metabolic disease called phenylketonuria (PKU). A child born with this disease will suffer catastrophic brain damage soon after birth unless his or her diet is low in phenylalanine and high in tyrosine. When fed the rather unpleasant and expensive PKU diet, those who tested positive for PKU develop normal intelligence. However, sticking to the special diet requires discipline and often leads to social disruption, especially during adolescence. It is a high price to pay for having an unlucky choice of parents.

When prenatal testing gives unequivocal evidence that the child has a 100 percent probability of a devastating disease for which there is no cure, those who are not following the scientific literature might think that the genetic defect should just be fixed. The promise of gene therapy looked good twenty years ago, and hopes were raised that hereditary diseases could be cured in utero. Progress slowed as numerous pitfalls were encountered. It is still not possible to just insert the gene for the enzyme that converts phenylalanine to tyrosine and cure PKU, or the CFTR gene and cure cystic fibrosis. When a fetus is found to carry Canavan disease, why can't the fetus just be given the missing gene for aspartoacylase? Because no one has found a safe way to insert a gene into a fetus or a newborn and keep it stable. Work toward this end continues, but there is a long way to go. For the moment, the most humane recourse is to abort the afflicted fetus.

GOOD GENES/BAD GENES

Smallpox and polio are close to being eradicated worldwide, and leprosy, guinea worm disease, and Chagas' disease are under the gun. If it is possible to eradicate bad bugs, why isn't it possible to eradicate bad genes? The problem is that most deleterious genes are recessive, and the trait inherited from one parent is carried harmlessly in the population as long as the copy inherited from the other parent is normal. Only when both copies are mutated will the individual be affected. The hereditary diseases that I have discussed here are so severe that the afflicted never have offspring. So from the point of view of population genetics, their births are not significant. What matters is how many people carry the trait. The parents of a fetus with beta-thalassemia each carry one normal copy and one abnormal copy of the globin gene. Random distribution of the copies predicts that three-quarters of their children will be normal, but two-thirds of these will be carriers. This will be the case whether or not the afflicted sibling is born. The original mutation accumulated in the population on Cyprus and other Mediterranean countries because the carriers were less susceptible to malaria. Now that malaria is almost unheard of in Europe and the Americas, there is not much use for carrying the beta-thalassemia gene. It would be nice to eradicate it.

The only way to get rid of a bad gene is to use the techniques for directed evolution discussed in the previous chapter. Most of them are totally unacceptable. Cloning of individuals who are free of the mutated gene could eradicate it, but the cloning would have to be on such a massive scale that tolerating the associated problems would be completely out of the question. Cyprus, for example, might end up being populated by a restricted number of clones and suffer from subsequent inbreeding. Replacing the mutated globin gene with a good copy in ES cells and then generating fetuses would work, but there are almost ten thousand carriers on Cyprus, and all of their children would have to be generated this way. Perhaps the best way would be to convince carriers not to have any babies. If they want the parenting experience, they could always adopt a child.

An infertile couple who have adopted their children are congratulated in public but often pitied in private. People have an inordinate fondness for their genes, or what they think of as their genes. People who like themselves hope to pass on all their good traits as well as their good looks. However, reproduction in humans takes two, and the partner's genes will show up as well. Moreover, as we have seen, many traits are recessive, and so the handsome man may easily carry genes for a goofy face or bowlegs. Since each sperm carries only one copy of each gene, which copy goes to the child depends on the luck of the draw. The same is true of the genes from the mother. Moreover, recombination between homologous chromosomes during the differentiation of sperm and eggs shuffles the genes, so that different combinations of maternal and paternal copies are found on the chromosomes of the germ cells. Each child inherits a set of genes different from those carried in either its mother or father, its grandmothers or grandfathers. It is not much of a surprise when a child is born with a trait that runs in the family, such as blue eyes or high blood pressure, even though both parents had brown eyes and normal blood pressure. Nowadays, it is possible to diagnose problems in reading and writing, such as dyslexia, and show that they are passed down through the generations, even though we do not know the gene(s) that might cause the underlying brain abnormalities. These problems are not so severe as to warrant considering genetic intervention.

When we start considering behavioral traits, such as being charming or patient, it becomes highly questionable whether they are the result of particular sets of genes. There probably are no genes for lying or greed, although the popular English writer on Provence, Peter Mayle, has written, "All lawyers have it. They can't help it; its in their genes." We are not just puppets jerked about any which way by our genes. Desires built up by a lifetime of deprivation or luxury can frame our actions independently of our genes. This is what the eighteenth-century philosopher David Hume considered to be free will.

There are also gray areas in behavioral traits. People have argued for years about whether left-handed people are different from the general population because of inherited genes, birth stress, or personal development. Is it nature or nurture? About 10 percent of people worldwide are left-handed. Centuries ago, lefties were considered sinister or gauche, and often shunned. Efforts were made to always hand the spoon to a child's right hand. But sometimes the child refused and took it with the left hand. Even now there is pressure to be right-handed in many cultures, just to fit in. For instance most scissors are designed to be used with the right hand; when used with the left hand, the blades separate and do not cut well. Recently, scissors built for left-handed people have been made available; in these, the blades are reversed so that they are pushed together when used with the left hand. Even with modern adaptations such as this, it is not clear how many righties might be closet lefties who were trained to fit in.

On the other hand, it is equally difficult to see how handedness can be an inherited trait like blue eyes. Identical twins are often discordant—in 18 percent of the cases, one is right-handed and the other is left-handed. Only 50 percent of the children in families where both the mother and the father are lefties are themselves lefties. Something is strange about this pattern of inheritance. The brilliant yeast geneticist Amar Klar has come up with a solution he calls the random-recessive model (Klar 2004). He proposes that there is a gene that in one form results in strong right-handedness. People who carry this dominant form of the gene will always use their right hand when throwing a ball, using a spoon, sawing, sewing, shooting marbles, bowling, cutting with a knife or scissors, hammering, and writing. Those who do not have the dominant gene have two copies of a recessive variant and their choice of hands will be random. This model accounts for the patterns of inheritance much better than any other model.

I remember sitting with Amar and my wife on our porch in California having a glass of wine when he explained the model. My wife mentioned that she was left-handed and all her sisters were right-handed. Amar sat up attentively when she said that her father was left-handed but that his identical twin was right-handed. It all fit. Amar was certain that my wife, like her left-handed father and right-handed uncle, carried two copies of the recessive form of the gene, while her mother carried both a recessive and a dominant form of the gene that he calls R. He told her there was a fifty-fifty chance that she could have developed as a righty, but that she happened to be a lefty.

The hypothetical recessive gene that results in a random event determining handedness has not been pinpointed yet, but Amar is pursuing evidence that it may be on chromosome 11. He hopes to show that segregation of one of the two strands of DNA at that position determines the fate of asymmetrically dividing cells during embryogenesis (Armakolas and Klar 2006). It would be exciting if he were right.

Does it matter whether you are right- or left-handed? The list of heroes throughout history who were left-handed is long, including Alexander the Great, Julius Caesar, Napoleon, Fidel Castro, Marilyn Monroe, Greta Garbo, W. C. Fields, Charlie Chaplin, Mark Twain, Leonardo da Vinci, Michelangelo, Picasso, Beethoven, Jim Watson, and Jimi Hendrix. They certainly did all right. Other lefties have not done so well—they have a much higher incidence of schizophrenia and bipolar disorders than right-handed people. Patients with serious mental disorders are three times more likely to be left-handed than the general population. This is not to say that everyone lacking the dominant R, or right-handedness, gene will go insane, just that they are at higher risk. Several other traits appear to be randomly established in the absence of the dominant R gene during early embryogenesis, which means that identical twins may not be truly identical. These include the direction of the hair whorl on the crown of the head and the selection of the brain hemisphere predominantly used for language function (Klar 2005; Weber et al. 2006). Most people have clockwise whorls, but about half of left-handed people have counterclockwise whorls. Likewise, most people predominantly use their left hemisphere in language tasks. However, those with counterclockwise hair whorls, who presumably carry two copies of the recessive form of the r gene, were often found to use both hemispheres, and 10 percent of them used the right hemisphere exclusively in language processing. It appears that the genetic determination of some traits is not always rigid but is given over to the consequences of chance in those individuals who do not carry the dominant R gene. (Disclaimer: I am right-handed but my hair whorl is counterclockwise.)

Finding a single gene "for" a behavior is the exception rather than the rule. A variety of behavioral traits, such as eating disorders, intelligence, addiction, risk taking, ability to dance, and homosexuality, have been followed in families in hopes of finding a genetic basis. The data are so messy that the geneticists usually conclude that these behaviors are controlled by a multitude of genes. A few candidate genes have been put forward, but it is far from clear whether the specific variants do more than correlate with the behaviors. And it is certain they do not work alone.

On the other hand, the flood of genomic information has allowed geneticists to recognize genes that correlate well with response to specific drugs given to alleviate or cure particular diseases. We are at the dawn of the age of personalized medicine. Cancers can be treated with a wide variety of drugs, but some patients respond to one and not another. Some patients require much higher than average doses to get any benefit. Choice of drug and dosage is usually determined by the physician carefully observing each individual patient until an effective regime is found. This is the "guess and test" approach to science and can be wearing on the patient until the proper treatment is found. It will soon be routine in affluent countries to characterize the pertinent genes as soon as an illness is diagnosed. Well-ordered medicine can be prescribed and the patient will benefit sooner.

Genetic information can indicate whether a person is especially sensitive to environmental toxins and hazards. When personalized genomic information becomes widely available, an individual will be able to make a well-informed decision to avoid working in a paint shop where toluene fumes are high. Others might not be bothered by this sweet-smelling organic liquid at the levels legally allowed, but the sensitive ones may become nauseous and, over time, suffer irreversible brain damage. By looking at the enzyme that modifies toluene so that it can be removed from the body, they will know whether they should stay out of the paint shop. If it is low or defective, they should also check the level of toluene in their drinking water. But what happens when the only shop in town uses toluene in large amounts? Should they pack up and move? Likewise, what happens when the owner of the shop has access to the genomic information of the applicants and refuses to hire anyone with even marginally low enzyme activity? Confluence of genomic and environmental information may one day constrain workers to jobs where there are no risks to them, but ones they don't want. Just as we need to have wide discussions concerning directed evolution, we need to have wide discussions concerning privacy and counseling on genomic information.

There is another problem not far down the road—the health insurance business as we know it may become obsolete (Kitcher 1996). When genomic information is available for the individual, it will undermine the basis for cooperation in a broad group. At present, the probabilities of illness or death are known for the group but not for the individual. This will change as we learn to interpret genomic information. The genetic basis for a considerable number of rare but devastating diseases is already established, and soon it will be possible to recognize the telltale signs for mildly debilitating hereditary diseases. When a person is known to carry such genes, insurance will be hard to get. Soon we may be able to scan a genome and assign risks for a large number of common diseases and disabilities. The insurance companies will get hold of the information one way or another and charge appropriately. A person carrying a gene that may double the chance of a heart attack before the age of forty will have to pay an exorbitant fee for life insurance. A person with a gene associated with slightly impaired immune responses may have trouble even finding a company that will provide health insurance.

In the past, the few who were unlucky in their genes were supported by the many who fit the actuarial tables in which the frequencies of things happening in the general population are collected. Since it was impossible to know beforehand who would be lucky and who would be unlucky, the same premiums were charged across the board and everyone benefited, including the insurance companies. When the odds are known for each individual, it changes everything. No company would insure a ship that is known to have holes in its bottom, but they will insure a sound ship against damage in a hurricane. The chances of being blown onto the rocks is small and unpredictable. For a price, they will cover it. However, if a crystal ball showed that a particular ship was more likely to encounter a storm at sea and founder, the underwriter would turn down the business. Genomic information can act as a crystal ball for health-related matters or workplace safety. While laws can be passed prohibiting the use of genomic information, in this day and age of computer files it is almost impossible to keep information private. Inevitably it will be used for discrimination.

It seems to me that the only solution is to do away with the sort of insurance policies that have grown up over the last few hundred years and replace them with a government-funded safety net for all. We can predict with some accuracy the frequency of individuals in a large general population who will inherit bad genes, and plan ahead for their support. By spreading the burden over the whole society, the unlucky will be spared and we can all feel more secure. A society that does not protect its weakest members cannot call itself civilized.

The maps and base sequences of the twenty-three human chromosomes present the complete genomic information for a human. The trouble is, we do not know how to interpret most of it. Only about 2 percent of the DNA codes for proteins, and the signals that determine where and when and at what levels the proteins will be found, have not yet been recognized. Moreover, a single protein never acts alone. Most function in complexes with other proteins, and the complexes form interconnected networks that are integrated between cells, and so on. To understand how humans work, it is not enough to know which proteins can be made. We have to think of the human body as a functioning system much like a complicated democracy with majority rule, vetoes, budget deficits, functionaries, and a lot of inefficiency. We have a lot to learn.

The conservative nature of evolution has allowed us to clearly recognize the few thousand genes that encode ancient universal proteins. These proteins carry out most of the basic metabolic processes that permit a cell to grow and divide. We've learned how they work from detailed studies in bacteria and model systems. Once we recognized that Canavan disease resulted from defects in the gene for aspartoacylase, it was not hard to figure out the biochemical defects. Likewise, recognizing that the chloride channel encoded by CFTR is abnormal in patients with cystic fibrosis helped to explain their breathing problems. It was no big surprise when hereditary anemias were traced to defective hemoglobin genes. Animals are all pretty much the same at the cellular level.

Some of the differences in lifestyles between animals are determined by the genes they inherit. Some fish are solitary, while others school together. Coyotes generally hunt alone, while wolves hunt in packs. Humans are social beings and feel uncomfortable if they cannot be with others from time to time. We know very little about the genes responsible for these feelings and behaviors, but some of the little that is known is presented in the next chapter.

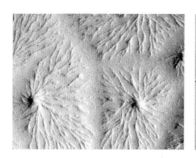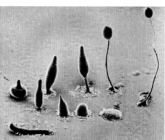

FIVE

SOCIOBIOLOGY

NATURALISTS AND EVOLUTIONARY BIOLOGISTS HAVE BEEN TRYING to explain the inheritance of cooperative behavior in social organisms ever since the problem stumped Darwin. The eminent entomologist E. O. Wilson made a forceful argument that William Hamilton's 1964 theory of inclusive fitness and kin selection could account in a Darwinian manner for much of the nepotism and aggression seen in social insects, fish, birds, and mammals. In many social species, individuals forgo having their own offspring in order to help other parents raise their young. Helpers in these species are very good at recognizing close relatives, who benefit from their attention. Since these relatives and the helpers have many of the same genes, the genes for such altruistic behavior are passed down through the generations.

In the last chapter of his book *Sociobiology*, Wilson extrapolated the concepts to human behavior, and that is when all the

---

FIGURE 5.0 *Dictyostelium discoideum*. Left-hand photo by the author, and right-hand photo by Larry Blanton.

trouble began (Kitcher 1984). Evolutionary scientists, human geneticists, social scientists, philosophers, and others were quick to argue that human social traits are such recent acquisitions that they could not have been shaped by evolutionary pressures; they could only be the result of cultural inheritance. These people rejected biological determinism as a political fallacy (Lewontin 1980). However, as long as we leave gender, class, racial, and individual human differences out of it, sociobiology has much to say about the underpinnings of basic, nonconscious behavior. Human cells and tissues are not much different from those in other animals. What sets us apart is that we are highly social beings who interact, plan ahead, and learn as we go. We are the sum of systems working at many levels. But we can learn a lot about humans by studying other cooperative, though simpler, model organisms.

SOCIAL GENES

The soil amoeba *Dictyostelium discoideum* is one of the simplest social systems and yet presents a wealth of social genes (Loomis 1975). When food is plentiful, each individual amoeba is on its own. It engulfs bacteria and yeast, grows, divides, and moves on. Only after the food source dwindles do the cells signal and respond to each other. Large numbers of amoebae then band together to make a slug shaped organism that can escape starvation by going somewhere else and waiting out the hard times. This species diverged from the evolutionary line leading to humans about 700 million years ago but still shares some of the same signaling mechanisms.

As soon as the food runs out, *Dictyostelium* cells express genes for making the small molecule cAMP and the surface receptor of cAMP. When cells start releasing cAMP, surrounding cells respond by moving toward the area where there are the most cells and the concentration is highest. The response is initiated when cAMP binds to receptor molecules on the surface and triggers changes inside the cell, which control their shape and movement. This chemotactic response results in hundreds of thousands of cells forming an aggregate, where they stick to each other and construct a covering over the whole aggregate. Cells that are genetically unable to make cAMP are completely asocial and just lie there when the food runs out. The same is true of cells that cannot make the cAMP receptor. These cells would not last long in the wilds; they would get eaten. Cells that can't make the enzyme that breaks down cAMP also fail to aggregate. In this case their problem is that they get swamped in cAMP and cannot tell where the highest concentration is.

Communication between cells using cAMP is different from communication between nerves and muscles, but the principles are the same. Animals' nerves release the small molecule acetylcholine, and their muscles respond when it binds to their acetylcholine receptors. They also release the enzyme that breaks down the signal, acetylcholinesterase, to turn off the signal. When this enzyme is inhibited, such as by a toxin in the venom of a green mamba snake, the muscle becomes paralyzed. Cobras have venom with an inhibitor aimed at the acetylcholine receptor. Their bite, too, paralyzes unwary prey. The genes that make acetylcholine receptors and acetylcholinesterase can be thought of as social genes, but they are really just the nuts and bolts of cellular communication. A vertebrate lacking these genes not only would be asocial but also would be dead. Communication between nerves and muscles is absolutely essential for many vital functions, such as making the heart beat.

Once *Dictyostelium* cells have aggregated and become mutually adhesive, they migrate around as a small sluglike organism and move up through the soil toward the light. When they get to the top, they release another small molecule, GABA (gamma-aminobutyric acid), which binds to a receptor specific for GABA (Anjard and Loomis 2006). This came as somewhat of a surprise to researchers, because GABA is one of the major neurotransmitters used in vertebrate brains to modulate neurotransmission, and its receptor was thought to be found only in animals. The *Dictyostelium* receptor was discovered only when the whole *Dictyostelium* genome was sequenced.

The fact that GABA is used as a signal in an organism that diverged hundreds of millions of years ago shows that this neurotransmitter has an ancient heritage as an intercellular signal. Drugs such as Valium modify how the brain responds to GABA and can make us drowsy. Moreover, Valium makes *Dictyostelium* cells become dormant spores. This drug mimics a peptide that controls the GABA response in vertebrates. In *Dictyostelium* an almost identical peptide is generated in response to GABA and then binds to another receptor on the surface of nearby cells, where it triggers spore formation. The circuitry is somewhat different, and the receptors are quite different, between mammals and the soil amoebae, but the conservation of signals over hundreds of millions of years is astounding. It looks as if once a good signal is found in one organism, it is retained in all the species that evolve from that organism. Mutants of *Dictyostelium* that cannot make GABA, or have lost the gene for GABA receptors, make very few spores. Moreover, they do not release the peptide that actually triggers spore formation. These genes, along with those that make the signal peptide and its receptor, can be thought of as social genes in that making spores is a communal process. Some cells in the little slugs do not make spores but build a stalk to lift the ball of spores up above the soil.

GABA is advertised as a diet supplement that reduces anxiety and increases mental clarity. Herbal medicines trumpet the benefits of GABA, but there is scant evidence it does anything before being metabolized. Very little ever gets to the brain. Ethanol is thought to affect some GABA receptors, and it certainly gets to the brain. Defects in GABA signaling are associated with epilepsy and Huntington's disease. Attempts have been made to deliver GABA to the brains of patients, but the results have been disappointing. In any case, these genes can be thought of as social genes, since debilitating symptoms arise when they go awry. But then, many genes that can result in diseases affect the behavior of the patient and are not all that different from genes that are necessary to make strong legs or strong minds. Almost all genes have some consequences for social behavior, even if only indirectly.

Another aspect of *Dictyostelium* has brought it into the field of sociobiology. It may be hard to think of a simple collection of identical amoeba as a social system, but remember that, as development proceeds, some cells build a stalk and do not become spores. These cells are giving up the prospect of seeding a new generation as they become trapped in the stalk and die. However, their sacrifice benefits the rest of the cells, since spores can be dispersed to new fields more easily if they are held up on top of a long, thin stalk. In most cases, the cells in an aggregate are all members of a clone that grew from a single spore. All the cells share the same genes, and so little is lost when stalks are made. It is not very different from a plant that devotes most of its cells to making the stem and a few cells to making flowers. The plant is all one organism, and the leaves are there just to help the flowers grow. Likewise, you don't worry about the fact that your heart and brain will die with you, and that only sperm or eggs have a chance to add your genes to the next generation. Your heart, brain, and the rest of your body altruistically help pass on the genes that are shared by every cell in your body. But if a friend helps you out, with no advantage to himself or herself, then we appreciate the altruistic behavior.

We would like everyone to be altruistic, but unfortunately there are cheaters. *Dictyostelium* also has cheaters (Strassmann, Zhu, and Queller 2000). When unrelated cells happen to come together and form a mound, sometimes one group "cheats" by not making their fair share of stalk cells. They make more than their fair share of spores. If this goes on for long, they should displace the losers. However, it has been found that most cheaters are not very good at developing on their own. The gene wars between cheaters and resistors are long over, and populations of *Dictyostelium* now live peacefully together most of the time. We know very little about the genes involved in determining which cells get to make spores and which ones do not, and so cannot say whether there is much to learn about human social genes by studying a soil amoeba. However, these quantitative studies with a microorganism have sharpened our thinking about what it takes to be social and how genes might affect it.

Many species of insects live in highly structured societies, and they have their cheaters as well. In some arrangements the most dominant individuals have a greater chance of passing on their genes than the more subordinate individuals. However, some less than ideal individuals take on the signals of dominance to increase their chance of having offspring. The tricks may include producing volatile signals or having the right facial markings, although they do not have the rest of the traits for dominance. These attempts to cheat are usually detected by the rest of the brood and seldom succeed. However, the genes for cheating get passed on, and the behavior persists year after year. The trouble is that we have no idea what the genes might be. Those who focus on the sociobiology of insects have a lot of ideas but little actual proof of how behavior is controlled or what the social genes might be. Nevertheless, it is clear that patterns of social interaction are inherited. The readout may be purely instinctive, but this does not make it any less important for the survival of the species. Cheating and catching cheaters is part of life.

Memory is another aspect of social behavior. *Dictyostelium* cells have memory only at the most primitive molecular level. A cell that has been developing for several hours accumulates new proteins that are not present in cells that are still eating. If they are all mixed together, the developing ones have a head start and will make more than their fair share of spores. Proteins last for several hours, and so developmental history matters. However, this sort of molecular memory is not what we mean when we talk of memory of time passed. We are much more interested in remembered places and times, memories of what we did and what others did. This sort of memory takes a brain, although surprisingly simple brains seem to be able to form and retrieve memories.

INVERTEBRATE BEHAVIOR

The worm *Caenorhabditis elegans* has a brain of sorts in a ring of nerves around its neck. There are only 302 nerve cells in each *C. elegans*, and many are spread out. Altogether the nerves make about four thousand contacts with each other. Modules of three or four cells in the nerve ring are linked up into networks that can process information and store it. Various neurotransmitters including GABA, acetylcholine, glutamate, and dopamine, are used to enrich the conversation between nerves. Corey Bargmann and her colleagues at Rockefeller University in New York showed that worms can learn the difference between good food and bad food (Zhang, Lu, and Bargmann 2005). They exposed worms to noxious bacteria for four hours and then gave them non-pathogenic bacteria to eat for another four hours. The worms

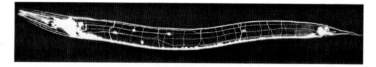

FIGURE 5.1 The nerves of *Caenorhabditis elegans*. The mouth is to the left and is richly enervated. The nerve ring lies just behind the head. A ventral nerve cord runs the length of this one-millimeter worm. The entire nervous system is expressing the gene GFP (white). Photo by Harald Hutter.

were then given a choice of either the bad bacteria or the good bacteria. Almost all of the worms chose the good bacteria, clearly demonstrating conditioned avoidance of chemical signals.

Chemical signals are also used by male worms to locate the vulva of a potential mate. The chemical signals are received by specific receptors on neurons at the male tail. Once the vulva is located, the male inserts his spicules and deposits sperm. Specific nerve cells control the extension of the spicules using acetylcholine as the excitatory neurotransmitter for this motor behavior. Twenty-eight genes for copulation have been found by searching through mutant strains that are mating-impaired. Many of these genes also play roles in other behaviors or neural functions and so are not dedicated to sex. Nevertheless, they certainly affect the social life of worms.

Insects have many more neurons than worms, most of which are concentrated in the head. Some species can carry out complex tasks that require learning and memory. There is a type of ant that forages over the flat, almost featureless terrain in the Sahara desert. The ants search near their nests for the carcasses of other insects, which they then carry back to the colony to be eaten. The outward journey can last up to an hour and cover half a kilometer. During this walk the ants make frequent turns and cross a lot of territory. Once they have a food particle, they carry it back to the nest in almost a straight line. This is quite a feat of navigation. It seems they are born with an internal template of the sky as seen in polarized light (Wehner 1997). They use the horizon to calibrate their skylight compass and then compare it to what they see as they move about. They inherently know about the movement of the sun during the day and bring this into the calculations. They then integrate their outward path and know how to go directly back to the underground nest.

In these ants the navigation data are collected and processed within a self-centered frame of reference rather than any external framework. An ant may not think about it, but its instincts are focused on itself. Dead reckoning gets them close to the small opening to the nest, but if they miss it, they use spatial clues in the vicinity of the nest. They can remember the arrangement of pebbles and hollows in the sand that they saw as they left the nest and integrate this knowledge with their sky map navigation to find the opening. These behaviors indicate relatively long-term memory coupled to a sophisticated computational mechanism that relies on global signals. They appear to assemble vector maps that they combine with route-based information and the distance traveled based on the number of steps taken (Wittlinger, Wehner, and Wolf 2006). If this fails, they initiate a search program that follows ever-increasing loops, starting and ending at the point they had thought was home. The longer an ant has been out foraging, the larger the size of the search loops it makes. This adaptation takes into account the accumulation of errors during longer forays. Getting home is a matter of life-or-death in the Sahara, and these ants have been selected to link together a set of subroutines adapted to their specific environment. Each subroutine has its limitations, but together they solve the problem. The ants compare their innate sky-map to what they see in the sky, as well as use their memory of spatial clues, and integrate it with their internal odometer. Clearly, these insects process a lot of information within their miniscule brains.

The sexual orientation of fruit flies has been found to be modified by mutations in the gene called *fruitless*. This gene is expressed in only about five hundred cells of the brain and makes an effective protein only in males (Kimura et al. 2005). Females do not have this protein, and these brain cells die. The presence of these male-specific nerves is essential for male-specific behavior such as courting, licking, and attempting copulation with females. These behaviors are abnormal or greatly reduced when *fruitless* males are put together with wild-type—that is, normal—virgins. Moreover, *fruitless* males court males and females indiscriminately. When *fruitless* males are grouped together, they form male-male courtship chains in which each male is both courting and being courted. The small network of brain cells seems to play a large role in the social life of flies.

RODENT BEHAVIOR

Mice are very social animals and will usually approach and get acquainted with a strange mouse that has entered their environment (cage). They will touch and lick the stranger, especially if it is younger. However, this behavior is missing in mice carrying defective copies of the *Fmr1* gene. These mice would rather spend time with an inanimate object than with a novel mouse. These mutant mice were generated from ES cells that had been modified by molecular genetic techniques that replaced their normal *Fmr1* gene with a mutated form. Mutations in *Fmr1* cause a form of mental retardation in humans known as fragile X syndrome, the most prevalent form of inherited mental disability. Many boys with this syndrome display autism reminiscent of the behavior of the mutant mice. They are dramatically asocial and severely impaired. We do not know exactly what is wrong with the brains of individuals with the *Fmr1* mutations, but they seem unable to establish appropriate social interactions. The *Fmr1* gene encodes a protein that modulates the levels of some mRNAs. However, this does not tell us which specific mRNAs are sensitive to the loss of *Fmr1*. Besides having cognitive defects, half the patients with fragile X syndrome have flat feet, large ears, and high-pitched speech. It seems that many processes go wrong in the absence of *Fmr1*. In Finland and Israel, pregnant women can be screened for *Fmr1* abnormalities, and those who are found to be carriers are offered prenatal diagnosis and abortion when it is called for. Participation in both the initial testing and prenatal diagnosis has been high, and many tragedies have been avoided.

When mice are placed in a novel environment, they run around the edges but are scared to enter the open space in the center. The molecular basis for this behavior is beginning to come to light after the study of mutants that are relatively fearless. Some fears are innate, such as the fear of heights, open spaces, and the shadow of a predator, while other fears are learned as the result of life experiences. The memory of fear is easily established and normally lasts for the duration of the animal's lifetime. A specific part of the brain, the lateral nucleus of the amygdala, plays a critical role in establishing memory specifically related to fear. As a first step toward unraveling the circuitry of fear, mice were generated from ES cells lacking amygdala-specific genes. Mice lacking one of these genes, stathmin, were found to have lost much of their fear of open spaces (Shumyatsky et al. 2005). When placed in an open field box, they did not seem to mind being out in the center.

The stathmin-mutant mice also seemed to have difficulty remembering averse experiences such as receiving a fairly mild shock when they were placed in a conditioning chamber. When wild-type mice were returned to the chamber the next day, they would freeze, remembering the painful experience. The mutant mice spent considerably less time frozen in place and soon started moving around. Their memory for other tasks was unaffected, and they showed no impairment of any brain functions other than the establishment of fear memories. These results show that processes in the amygdala, one of the central clearinghouses in the brain, are essential for both innate and learned fear, an emotion essential for survival and social behavior. Perhaps we are beginning to recognize a few genes for basic behaviors and emotions, but we should not assume that more complicated social interactions are all genetically controlled.

The social behaviors of ants, bees, and other insects have been intensively studied for the last hundred years and have yielded amazing insights. Kin selection has been accepted as the explanation for many cases of altruistic behavior in which an individual sacrifices its own benefit for the good of siblings. If this increases the chances for the passage of genes shared with close kin, then the individual sacrifice benefits the species. Likewise, the drawbacks of internecine competition become clear when cooperating groups invade and replace feuding native species. The temptation to extrapolate insect sociobiology to human societies is undeniable—but probably faulty because much of human behavior is learned. Even extrapolating from asocial mice and fearless mice to autistic humans and Rambo-type humans can be misleading, because we humans continuously modify our behavior according to conscious decisions.

During the first half of the twentieth century, researchers made an enormous effort to purify individual proteins and learn their properties. During the second half of the century, researchers made a similar effort to recognize the genes encoding these proteins and learn about their regulation. Such reductionist methods were highly successful in defining the parts lists of cells, but not very successful in explaining the properties of different cell types or predicting the consequences of varying one or more of a cell's components. It became clear that an individual is not just a collection of genes. In even the simplest cells, genes can provide only proteins, which must then interact in integrative networks to carry out the business of the cell. When considering organisms made up of different cell types, the story gets only more complicated, since the unit of selection is the organism, not the cell. Holistic approaches are required if we are to grasp even the rudimentary fundamentals of a living organism. While the techniques of biochemistry and genetics have been successfully applied to complex processes such as embryogenesis, the results have had to be interpreted at multiple levels, from the molecular to the cellular to the organismic. Systems approaches that used the wealth of information available at the molecular level and integrated it into simulations of metabolism, cell growth and division, and intercellular communication provided a new way to look at organisms.

All multicellular organisms have a variety of specific receptor proteins on their surfaces that respond to external signals and affect cellular responses. The GABA receptor on the surface of *Dictyostelium* cells is very similar to the GABA receptor on the surface of human brain cells. The mechanism that allows our brains to respond to GABA has not changed significantly in a billion years. But what happens next in the brain depends on the particular species and particular cell type. *Dictyostelium* does not get epilepsy when GABA signaling is defective. Human brains not only depend on the ability to make and respond to GABA, but they also must carefully adapt their response to the state of the particular neuron in a particular part of the brain.

We respond to bad food by forming an aversion, just as the simple worms do. The same neurotransmitter is used, but the wiring is completely different. In *C. elegans* the information that potentially noxious bacteria have been eaten is registered by a single pair of nerves that respond to serotonin. This minimal network is enough for conditioned aversion in the worm. People too transmit gustatory signals by serotonin, but the serotonin-sensitive cells in the brain number in the billions. The circuitry is certainly different from that in worms. Bad food leads to emotions of disgust and even fear, which wend their way around the brain to finally result in the behavior that makes us say, "No, thank you, I am afraid I don't eat that."

The surprising similarity between the abnormal social behavior of mice carrying defective copies of the *Fmr1* gene, and that of people in which the same gene is mutated in fragile X syndrome, indicates that some of the neural functions of the *Fmr1* protein may have been conserved from mice to men. It doesn't say that the brain defects are exactly the same, only that we should try to use mice to find the critical targets of the *Fmr1* protein in mouse neurons, because this could lead to a better understanding of devastating human mental disabilities. Optimists would say that drug studies in *Fmr1* mutant mice may uncover better treatment for affected humans. Some basic central nervous system processes in mammals may have changed little in 50 million years of evolution.

The genes affecting sexual orientation in flies are unlikely to be relevant to behavior in humans. However, they do show that complex behaviors have a genetic basis in simple organisms driven only by instincts. Choice of partners in human society is clearly the composite of inputs at many levels, and the instinctive portion is probably quite low.

Stathmin is one of the ancient conserved proteins. The amino acid sequences in mice and humans are 95 percent identical. Although essential for innate and learned fear in mice, stathmin may not be required to put fear in the minds of humans. It plays a general role in controlling cytoskeletal elements within cells, and this might have led to its conservation in vertebrates independent of any role in behavior. Its expression pattern in the lateral nucleus of the amygdala may be specific to mice. We might or might not use similar neural networks in generating our feelings of fear. We do have inbuilt fear responses, but they can often be overcome by rational thought.

We are descended from organisms that gave rise to flies and mice as well as apes, and we inherited many of the genes selected at those stages of evolution to cope with the challenges faced at the time. Other genes were lost when their roles were no longer pertinent to the new lifestyles of the intermediates. The basic social processes are still at work in us; we simply added many more.

Perhaps our biggest addition was language. We are the only species able to discuss abstract concepts and private feelings with ourselves and others. Language is the basis of declarative memory and the foundation for higher consciousness. However, the evolutionary steps that led to grammatically correct speech are far from clear. Many small anatomical changes apparently were necessary to produce the required range of sounds, but so far only one gene, FOXP2 on chromosome 7, is directly linked to language. This gene encodes a transcription factor that evolved in the human line, permitting us to communicate more clearly. Several family lines carry mutations in FOXP2 that result in severe language and grammar difficulties. Most likely, this mutation affects the expression patterns of several genes in the afflicted members of this family. The rest of us take language for granted and use it to define ourselves and those around us. These processes are the focus of the following chapter. When we define others, we judge them. Some are nice and some are not. We spend considerable effort finding out who is getting a free ride and punishing the cheaters. In a later chapter, I discuss the advantages of such learned behavior in human societies.

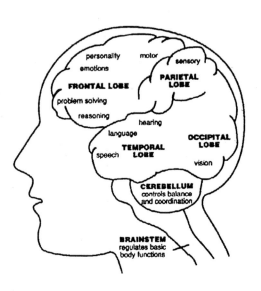

personality

motor

sensory

emotions

PARIETAL LOBE

FRONTAL LOBE

problem solving

reasoning

hearing

language

OCCIPITAL LOBE

TEMPORAL LOBE

speech

vision

CEREBELLUM
controls balance
and coordination

BRAINSTEM
regulates basic
body functions

SIX

CONSCIOUSNESS

IT APPEARS THAT ONLY HUMANS ARE FULLY CONSCIOUS OF BEING alive. We are aware that we are aware, and we feel there is something special within us that makes us who we are. The soul or spirit is considered by many to be central to our being, our life principle. But we don't know what it is. The study of consciousness has only recently become scientifically acceptable as the result of major strides in understanding the brain. Using modern imaging techniques, we can now peer into the mind as emotions, feelings, and responses pour out while a subject listens to music, makes a decision, gazes at a scene, or hears joyful news (Moll et al. 2005; Hsu et al. 2005). The neuroanatomy of fear, disgust, anger, sadness, surprise, happiness, anguish, pride, and other emotions is becoming better known, but we still cannot fulfill the Delphic injunction carved into the lintel of the Temple of Apollo, "Know thyself."

FIGURE 6.0 Brain. Courtesy of the South Carolina Department of Disabilities and Special Needs.

Francis Crick was the first to challenge me to think quantitatively about consciousness. For over thirty years Crick tried to formulate a concept of consciousness that could be tested scientifically. Ever since deducing the double-helix structure of DNA with Jim Watson back in 1953, Crick had been able to pursue an unconstrained career in theoretical biology that lesser thinkers could only dream of. In 1976 he moved to the town where I work, and we saw each other at times over the years. After solving the molecular basis of heredity, he wanted to work in another field that was just as challenging and important—the mind. From 1976 to his death in 2004, he grappled with the meaning of consciousness, often in collaboration with Christof Koch, a professor of neurosciences at Caltech.

In 1997 Crick gave a lecture at the University of California, San Diego, with the title of "Consciousness: The Nature of the Problem." He started off by saying that he would talk about consciousness but that no one should ask him to define it. I remember almost walking out of the hall, because this could not be science. Only my respect for Crick kept me in my seat. As I listened to what he had to say, I tried to formulate my own definition but failed. The best I could do was consider a metric for consciousness. We all agree that a stone or a dead animal has no consciousness, and we could probably agree that a wide-awake, animated, smoothly talking Francis Crick was about as conscious as could be. So my scale for consciousness is measured in centi-Cricks, with Francis at the top with one hundred centi-Cricks, and the stone at the bottom with zero centi-Cricks. A sleeping person is not completely unconscious, especially during dream-sleep. What do we give a dreamer, five centi-Cricks or fifteen centi-Cricks? At least we have started a kind of quantitative discussion.

Antonio Damasio is a highly respected neurologist who has come up with clear ideas concerning consciousness based on years

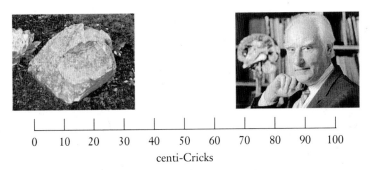

FIGURE 6.1 A scale for consciousness. A stone has no consciousness, while a wide-awake Francis Crick registers one hundred centi-Cricks. Left-hand photo courtesy of Francisco Reinking. Right-hand photo copyright Salk Institute for Biological Studies. Photograph of Francis Crick used by permission of Mrs. Francis Crick and the Salk Institute for Biological Studies.

of observing brain-injured patients in the clinic. He has suggested that consciousness can be divided into core consciousness and extended consciousness. He thinks of consciousness in a graded manner that can be either more or less impaired depending on the specific brain injuries. "My view is that both consciousness and attention occur in levels and grades, they are not monoliths, and they influence each other in a sort of upward spiral," he writes in *The Feeling of What Happens* (1999). I would be interested to know how he might score his patients in centi-Cricks.

Most of the information presented to the brain by the senses is ignored, not further processed. However, when illuminated by attention, sensory information is sorted, consolidated, and often symbolically labeled, such as by adding the quale of red to visual input from retinal cells that respond to long-wavelength light, or pain to the prick of a thorn (Damasio 1999; Koch 2003). Most organisms with a nervous system process sensory information to this level and achieve an awareness of the environment that serves their needs, but it is all unconscious. Core consciousness results from the confluence of awareness, emotion, and memory. It does not require all aspects of emotion or memory but needs some to begin to register on the centi-Crick scale. When projected onto memory of self, processed sensory information can lead to extended consciousness. Perception is continuously updated and modified by changes in both external and internal signaling to generate a stream of consciousness. When verbally presented to the brain, it allows executive functions such as planning and abstraction that characterize higher consciousness. I would say we are now above the halfway mark on the centi-Crick scale.

Awareness that allows us to register sensations is essential for consciousness. So too is the sense of self that comes from the unconscious signals that the body continuously sends to the brain. The rumbles of the stomach, the beating of the heart, the temperature of the blood, and the position of the limbs are some of the myriad neuronal signals that reach the brain stem and tell us we are there. We are continuously informed by gravity of what's up and what's down. Visceral emotions inform us of well-being or ill-being. We usually do not attend to them; the body does. They are background feelings and emotions that color our lives. They affect posture, speed of motions, and tone of voice. Fatigue, excitement, tension, harmony, and discord can affect background emotions and stimulate or inhibit our drives, our sense of purpose, and our moods. With attention and emotion, we can generate core consciousness even if we have little or no feeling of it. Plans for movement are made before movements are made, just as responses occur before emotions are formed. But the difference may be only a fraction of a second.

Extended consciousness appears to depend on core consciousness and feelings that enter into memory, where they can be pondered. An autobiography is built up over a lifetime and kept in memory for future reference. Language allows us to discuss the internal images, the movie-in-the-brain, that represent our consciousness. With language we can formulate abstract concepts and share our private feelings with others. We will never be able to enter the consciousness of another, because our own consciousness intervenes, but I will try to tell you my thoughts and hope you will tell me yours.

In Damasio's view, core consciousness must be continuously created by inputs from both inside and outside the individual. It may be generated many times within a single second and is continuously informed by signals from the autobiographical self that is held in memories. However, core consciousness does not depend on lasting, autobiographical memory. Damasio describes one of his patients who had completely lost the ability to remember anything for more than a minute. At age forty-six, this patient, whom Damasio calls David, developed viral encephalitis, which destroyed large parts of both hemispheres of his brain, including the region of the brain critical for establishing memories, called the hippocampus. The man was unable to remember his name or learn any new facts, but he walked, talked, ate, and drank quite well. He was polite, but he had no memory of his wife or children or any other individual he once knew. His short-term memory was unimpaired, so he could get around and recall what was said a few seconds ago. A minute later he had no idea what just went on. He lived totally in the present, yet his core consciousness was intact. He was awake, alert, and responsive when he got up in the morning. He could be attentive and focus on a task. He could even play checkers when he felt like it and certainly seemed to enjoy winning. But he had no idea of the name of the game. He seemed to be able to form emotions without any benefit of conscious memories. Some would consider him fully conscious, but his impaired memory meant that his autobiographical self was never updated, and he had lost extended consciousness of his previous self. If you spoke with him for a few minutes, you would not consider him normal. You would probably conclude that he was not all there.

Within his window of short-term memory lasting forty-five seconds or so, David's sensory inputs and responses were his own, not somebody else's. Core consciousness can exist as a stream of consciousness with no yesterday and no tomorrow. For David, individuals and specific things had no particular importance. He could not say why particular places or people pleased or annoyed him, just that he was pleased or annoyed. While his massive brain damage left him with little or no extended consciousness, his core consciousness worked well with only the remaining basal structures in the hindbrain. When comparing a wide-awake David to a person in a coma or a person in a persistent vegetative state, I suggest that he might have twenty-five centi-Cricks of consciousness, but without many more such cases we really don't know how to set the scale.

ANIMAL BEHAVIOR

As usual when approaching an almost impossibly complicated problem, I like to start at the simplest, lowest level and then try to work my way up. Although no one has ever suggested that an amoeba or bacterium is conscious, single-celled organisms have properties of responsiveness and purposeful behavior that are necessary but no way near sufficient for consciousness. Consider that the bacterium *E. coli* and the amoeba *Dictyostelium* are able to respond to specific chemicals in the environment and move in the direction of the highest concentration. These cells are not asleep or dead. They are showing adaptive behavior in relation to the source of food or the cells with whom they can cooperate. But that is about as far as it goes for information processing. Similar cellular responsiveness is required for awareness, but it has to be coupled to a central nervous system to form the under-pinnings of even core consciousness.

Worms and flies show no signs of consciousness, but they do appear to have a limited sense of self. The male worm moves purposefully when attempting to copulate and knows his head from his tail. It seems likely that flies have visceral emotions that come from the state of their bodies. They can sense when they have eaten their fill; they remember having been shocked and show fear when returned to the same place; they court in stereo-typed manners in response to pheromones. However, they have the attention span of a fly and show no indication of even the beginnings of core consciousness or emotions, let alone the feel-ing of emotion. I would not give a wide-awake fly even a tenth of a centi-Crick.

The navigational skills of ants in the Sahara, salmon returning from the sea to spawn in the rivers they were hatched in, and the seasonal migrations of birds all emphasize the power of instincts. These animals have highly evolved neural networks that gener-ate innate responses to spatial clues. Their rudimentary form of intelligence gets them where they need to go, but it is inflexible and cannot be directed internally to generate a feeling of what is happening. They do not register on the centi-Crick scale of consciousness.

The apparent intelligence of border collies herding sheep is phenomenal. At a whistle from the shepherd, these dogs will round up and move dozens of sheep from one pasture to the next without losing a stray. Then they lie down and watch. Although their fore-sight into the movement of the flock appears to result from careful consideration, they are just expressing wolflike instincts that have been selected in the breed for generations. A border collie at home will feign happiness to get a treat or sadness to gain affection. Dogs are consummate con artists and have learned behaviors that get their owners to do what they want. They are smart, loyal, and affectionate, but it is not clear if their emotions, feelings, memories, and thoughts are like ours. Man's best friend certainly has core consciousness of a sort, but does he have extended conscious? Should we consider the possibility of five centi-Cricks?

What about our closest living relatives, the chimpanzees: do they have extended consciousness? Are they aware that they are aware? They show their emotions by smiling or frowning much as we do, so there is little question when they are happy or sad. And they seem to know it as well. They express hurt feelings after losing an encounter with a dominant member of the tribe, showing that they can perceive their own emotional state and can project their emotions to others around them.

Chimps have excellent memories and a good sense of time. As youngsters they can learn sophisticated uses of tools by watching their parents and cousins. They vocalize in meaningful hoots and growls, but the shape of their larynx (voice box) does not allow them to speak as we do. Nevertheless, when raised by humans, they can acquire a vocabulary of dozens of words in sign language and apply them to novel situations. They may not have true language, but they have all the other attributes of extended consciousness (Seyfarth and Cheney 1997). Until we can communicate better with chimpanzees, I give them a tentative ten centi-Cricks.

Human consciousness may have increased dramatically as the result of competition within the species for territory, resources, and mates. Being aware of your own feelings allows you to guess the feelings of others and predict how they may act. We go on the assumption that core emotions and feelings are similar in others.

Introspection gives us information about our own private mental experiences of emotion, which can indicate how others may feel. When we see someone else in emotional distress, we often feel distress ourselves. Such innate empathy can lead to cooperative behavior in which weaker members are cared for (Gazzaniga 2005). Harmony and cooperation within a tribe increases as internal strife is kept to a minimum.

From our own experiences we can guess how others will react to threat or appeasement. Our predictions are not always right, but they are better than nothing. When members of another tribe invade, they can be repulsed more easily if you know how to scare them. One way to guess their reactions is to know how you feel and how you would react to shows of force or bluff. Knowing your own emotional responses can lead to more appropriate decisions.

When members of a tribe learned to describe their feelings and thoughts while sitting around a campfire, they could form a group-consciousness that was more effective than the experiences of any one member. The wisdom of the tribe could be applied to the problems that arose within it or when it encountered competing tribes. Talking was not all about where to find food and what to bring to the cave. It had to encompass vague emotions and confused feelings. We really have no idea how language evolved, because the line of great apes that finally led to *Homo sapiens* lost all its early intermediates. *Homo habilis, Homo erectus,* and *Homo neanderthalensis* are all extinct. The brain size of *H. habilis* was about half that of modern humans, which probably limited the ability to communicate subtle thoughts. Members of the *H. habilis* species lived in Africa 2.2 million years ago and disappeared about 1.6 million years ago. During that half million years, there may have been many improvements in language, but they left no record.

Fossils show that, about a million years ago, *H. erectus* appeared in Africa, China, and many places around the world. Members of this species had large brains, almost as big as modern brains, walked upright, and were about our size. Their obvious success as a species may have resulted from being able to communicate better than any other species. However, they had relatively flat larynxes, which would have limited the range of sounds they could have made. By our standards, their language skills may have been rudimentary, but we will never know. Later, the Neanderthals moved into Europe and thrived in the cold climate until thirty thousand years ago. They controlled fire, made spears, buried their dead in a ritual manner, and constructed complex shelters. A primitive flute found associated with Neanderthal remains in Divje Babe, Slovenia, that are about sixty thousand years old suggests that by this time they enjoyed music and may even have accompanied it with song.

A fossilized Neanderthal hyoid bone nearly identical to that of a modern human was discovered at the Kebara Cave in Israel (Arensburg et al. 1989). This floating U-shaped bone determines the shape of the larynx and holds the root of the tongue in place. Its size and shape indicate that the Neanderthals could have spoken as clearly and precisely as you and I, although their voices would have been high-pitched and somewhat nasal. Their brains were slightly larger than ours, but it is not clear whether their language was as rich and subtle as ours. Although they lived side-by-side with modern humans for a few thousand years, there seems to have been little interaction. Comparisons of DNA sequences from modern humans and Neanderthals show no evidence for mingling of the species. Now they are extinct.

Language may have developed to a high level in modern humans while they told stories to the tribe. Allusion, metaphor, and simile enhance a story. By embedding a phrase within a phrase or a clause within a clause, a storyteller can pick the appropriate level of precision. It makes a difference whether you tell the tribe that the region has animals that you can eat or animals that can eat you. For centuries, language has been refined to make laws, draw up deals, and outfox opponents. Good storytellers may or may not have had more children but certainly had better students.

When our senses record an event, it enters in a nonverbal manner but is rapidly converted into verbal presentation when we choose from memory the most likely explanation. Before we know the name, we know the thing, but the named thing is what we remember. We cannot willfully stop verbalizing an experience, and so it is mostly verbal images that enter into memory. Perhaps some of our internal movies are silent, but most of them are talkies. As we become aware of the internalized memory, our feelings may change and the story may gain additional emotional coloring. When the verbal image is retrieved, it may not be the same as the original experience, and the story may be altered by some poetic license. Just because we say something is so, does not mean it is so (this story included).

THE BRAIN

The brain is the seat of the mind. It is enclosed by the skull and connected to the world by various sensory organs, such as the eyes and ears, as well as by the spinal cord that runs down the back. On the few times that I have looked inside the skulls of mice, I have been overwhelmed by the complexity and beauty of the brain. I wouldn't know where to begin to try to understand it. Even neurobiology specialists are in awe of this organ that goes so far beyond simple analogies to computers. There are 100 billion cells, each connected to hundreds, even thousands, of other cells that can work in parallel or in series. Signals are transmitted along axons at almost lightning speed, and responses are measured in milliseconds. Yet modern techniques can now image the living brain and determine with considerable precision which areas are active at any given time. The imaging technique called fMRI uses magnetic resonance to measure blood flow that accompanies neural activity in the brain. Thimble-sized regions can be distinguished, giving good anatomical resolution. Slowly, the hidden workings of the brain are being discovered.

Studies on isolated sets of nerves have shown that some are excitatory and others inhibitory. They use different chemical signals, which are received by specialized receptors on the surfaces of other nerves (Rose 1992). In some cases an excitatory nerve is connected to two other nerves, which are themselves connected to each other. If one of them is an inhibitory neuron, then the other will be excited for only a few milliseconds before it is inhibited by its partner. Such feed-forward inhibition keeps the signal sharply delineated in time as it courses through the brain.

The specialists can recognize anatomical landmarks that characterize the human brain so well that they know exactly where things are happening. There are specialized lobes of the brain at the front, the sides, and the back of the skull that each have names related to their positions. The frontal lobe is the crowning glory of the human brain, having expanded more than threefold during primate evolution. But keeping it all straight can be difficult unless you have a mental map of the whole brain. Usually we can rely on the neurologists and neurobiologists to determine the details while we ponder how it all works.

There is a structure in the middle of the brain called the thalamus that appears to be critical to awareness. If it is destroyed by a stroke or other cause, the patient loses all awareness of self and the surroundings. Core consciousness disappears along with any hope of extended consciousness. The thalamus appears to be a central switching station that receives input from every sensory system, processes the information, and passes it along to the cerebral cortex. There are also more specialized regions, such as Broca's area in the left temporal and parietal lobes, recognized by the French neurosurgeon Paul Broca to be essential for speaking and understanding languages. When fMRI scans are carried out while a conversation is under way, they show activity in this specific area. Moreover, the regions that are stimulated by activity in Broca's area also light up. They include a region in the midline that is shaped like a seahorse (genus: *Hippocampus*). Memories of all sorts, including conversations, pass through the hippocampus (Rose 1992).

Damasio tells of a patient he called Emily, who suffered bilateral damage in the early visual cortices at the junction of the occipital and temporal lobes (Damasio 1999). She could see perfectly well and recognize everything around her, but she could not tell one person from the next. She had lost all associative memory relating faces to people. Emily did not even recognize her own face in the mirror. When shown pictures of her daughter together with pictures of strangers, she tried hard to pick out the one she loved but was totally unable to recognize her face. She knew what a face was but not to whom it belonged. Although fully conscious of her disability, she couldn't do anything about it. Her memory of visual images, specifically faces, was gone. This patient presents us with evidence for a specific region of the brain that is essential for associative memory of facial images. However, it does not mean that we have a portrait gallery of friends and acquaintances at the junction of the occipital and temporal lobes, only that circuits in this area play an important role in recognizing people by their faces.

The conclusion that the early visual cortices are involved in recognizing faces has been independently supported by direct measurement of brain activity using fMRI. When subjects are shown a face, their brains light up in exactly the regions that suffered damage in Emily. This experiment has even been carried out with unconscious patients who were in a persistent vegetative state. The same regions lit up when they were shown pictures of faces but the signal stopped there. It seems that this part of the brain can function independently of core consciousness, feelings, or memory. However, it needs many other parts of the brain to process the information into a useful form.

The fact that we have specialized neural networks for face recognition should not be too surprising, considering the selective pressure to distinguish members of the tribe from invading strangers that was strong even before our ancestors came down from the trees. One look at a stranger tells us that he or she may be a potential enemy. Mistakes can be costly, and decisions have to be made fast. We seldom mistake a stranger for an acquaintance and are surprisingly good at never forgetting a face. Perhaps this is why there was considerable discussion about a partial face transplant recently carried out in France. A woman whose face was so badly disfigured that it could not be repaired by plastic surgery received the face of a recently deceased woman. The two women were about the same age, size, and general appearance, but no one would confuse one for the other. After the face transplant, the patient did not look like her former self, nor did she look like the donor. Nevertheless, she looked much better than after her nose, lips and chin had been torn off. Within a few weeks her friends and family got used to her new appearance and changed their image of her. Some ethicists questioned the use of advanced skills in transplantation to treat the woman. They suggested that her identity had come under the knife, and that somehow she would take on the emotional and mental traits of the donor. Just because we recognize people primarily by their eyes, nose, lips, and cheekbones does not mean that these variables define them. It is simply how we know them. The transplanted tissue was only on the outside of the skull; inside, her mind was her own.

Just as we are born with a built-in ability to focus on faces, we are born with preferences for, and strong aversions, to certain tastes. Given a cup of sugar water, a baby will drink it up. Given a cup of saltwater, the baby will stop drinking, possibly spit it out, and pucker up its face. Adults react the same way to sweet or salty water. An exception to this response was found in a old man, referred to as subject B, who had suffered irreversible damage to large parts of his brain as the result of viral encephalitis that left him with complete amnesia (Adolphs et al. 2005). Like the patient referred to as David, who was discussed earlier, he could remember things only for about forty-five seconds and failed to recognize familiar people. He could form no new memories and had trouble retrieving many old memories. He had lost all ability to recognize sweet from salty and declared each to be "delicious." He drank cups of saltwater as eagerly as sugar water and always looked pleased. Comparison subjects, some of whom had extensive brain damage, stopped drinking the salt water after the first sip. When presented with cups of sugar water, they always drank it down and recognized it as sweet.

MRI scans of subject B showed that the portions of the brain involved in higher taste processing were damaged. MRI scans of the comparison subjects showed these areas to be intact. Amazingly, patient B could still discriminate between the tastes when given the chance. Although he could not recognize sugar or salt, when he sipped from a set of color-coded cups, half of which were filled with sugar water and half of which were filled with saltwater, he invariably chose the sugar water for the second sip. If the first was salty, he picked the other color. If the first was sweet, he stayed with that color cup. His case demonstrates the dissociation between behavioral preference and cognition. Decisions can still be made when brain regions responsible for preference of specific foods are compromised and there is no memory of what tastes good.

Other regions of the brain are essential for the recognition of melodies, tunes, and the inflections of individual voices. There is a tragic story about an opera singer who suffered damage in the auditory area and lost all ability to appreciate music or the work of specific musicians. He could hear perfectly well but could not associate what he heard with memories he had formed over his lifetime before the illness robbed him of his knowledge and talent.

THE SELF

I remember a spring morning bicycling along a country road when I first realized that I was aware of myself. I was only nine years old, but I knew it was important that I had a sense of self knowing self. For years I had been aware that I felt alive when I woke up in the morning, and that every night as I went to sleep my consciousness faded. But it had never occurred to me to be surprised that it was the same me—the same self—who woke up every morning. My autobiographical self just incorporated its author. I really had no idea what being conscious of self meant, but I thought it might have to do with being an individual— almost a grown-up. Many years later I learned that others had been grappling with the problem of what is self and what is consciousness for centuries.

Aristotle, the Greek philosopher who wrote in the fourth century B.C., was aware that there was more to living beings than just their bodies. The body was essential and informed the mind, but something immaterial was formed in the mind. He called it the soul and compared it to the flame on a candle. The wax and wick were essential, but without the flame, the candle gave no light. The soul allowed the individual to perceive itself and become alive. In his book *De Anima*, Aristotle approaches the soul from many angles but never quite defines it. Over the next fifteen hundred years, religions around the world emphasized the uniqueness of the soul and offered ways for its salvation. In 1996 Pope John Paul II said, "While man may have developed through an evolutionary process, only God can create a human soul." The devout were told that they were responsible for their souls and should live in prescribed ways.

In the seventeenth century, René Descartes focused on the individual self. This led to his famous conclusion: "I think, therefore I am." He was aware that he was aware. He was himself purely in his mind. Damasio, in a book entitled *Descartes' Error*, has called into question the way that Descartes came to his definition of self, but not the feeling of self. Modern neurophysiology has shown that the mind does not exist without the brain, and that the feeling of self is generated in the mind of each individual.

In the eighteenth century, Immanuel Kant separated consciousness of self into two forms: inner self-consciousness (the "I" of reflection) and outer self-consciousness (the "I" of doing). This duality can be thought of as the self being "partly object and partly subject," as the philosopher William James put it a hundred years later, as in "I was thinking to myself the other day . . ." However, Kant doubted that we could ever know ourselves as we are, only how we think we are. Advances in neurobiology are beginning to objectively define the limits of self and even unravel where the feeling comes from. Although Kant was burdened by the fear of death and dissolution of the self, his writings opened wider the doors of the Enlightenment and formed the underpinnings of modern cognitive science.

As we have seen, it is becoming clear that private mental experiences of emotions generated by signals from the brain and the body give rise to feelings at the threshold of being and knowing. When connected to memories, these feelings lead to extended consciousness that includes a sense of self. Information can be processed that includes the present and future self and can lead to appropriate decisions. Possible actions are considered, compared, and decided on. Within a split second, we respond. Reactions themselves are stored in memory, where they can form the basis for later decisions. The self is continuously updating its autobiography as it goes through life. The self allows us to enter our own movie-within-the-brain.

There is a pathological state called petit mal, or epileptic absence, in which the patient appears to temporarily lose all sense of self. The arms and legs still work and balance can be maintained, but the patient has no idea what is going on or what to do next. These episodes usually last only about twenty to thirty seconds and appear as brief staring spells. Patients are completely unresponsive during an attack and show no emotion. After it has passed, they are unaware that a seizure occurred. During the episode, the person is not in control, and motor functions are on autopilot. As long as the patient does not happen to be riding a bicycle or driving a car, he or she will probably be fine until the next attack. For a brief period the person is unconscious, standing there staring, awake and attentive, but unconscious. The self is absent.

The feeling of self is strong and well in most people. The sense of personal existence awakes with us each morning and drives our emotions and behaviors. Our sense of personal being gives us personal responsibility for our actions. We feel that the decisions we make are our decisions. Should I risk injury to help a stranger? Should I return the extra dollar I was given in change? We act with the feeling that rewards and punishments depend on our choices. However, in the back of our minds we know that we are controlled by previous experiences—being caught with our hand in the cookie jar, being shoved aside in a crowd, a million things. Is everything we do determined, or do we have free will?

When we consider the world around us, we see events unrolling in an orderly manner. Turning the handle allows the door to open. This is a case of simple cause and effect. Our hand grasps the handle, muscles turn the hand, the latch is withdrawn, and the door opens. Without these actions the door would not have opened. We take it for granted that turning the handle causes the door to open. In fact, without cause and effect nothing would make any sense. When a billiard ball is set in motion, we can predict the angle at which it will bounce off the side. When it hits another ball, we can predict the motion of both balls. If we are good at pocket billiards, the target ball will fall in a pocket. We depend on dependability. We know the Law of Conservation of Momentum and can calculate angles fairly accurately. If the cue ball disappeared when it was hit and then reappeared behind the target ball, we would give up the game. In fact, we would never play billiards or any other game if we did not think cause and effect controlled the world around us. We might continue to move, eat, and drink instinctively, but we would never plan ahead or hope to learn from experience.

Only because events appear to be determined by prior causes do we trust what our senses tell us. We may not know all the causes, but we trust that together they determine the effects. Every day our faith in determination is reinforced by our ability to predict the outcome of simple cause-effect relationships. However, this faith comes in conflict with our feeling of being captain of our own ship, the feeling that we may go as we please, avoid rocks, and come to a safe haven. What makes us so special that we are outside the web of cause and effect? Where do we get the privilege to do as we like and not be puppets controlled by prior events? Unfortunately, there seems to be no clear answer to this question.

We are comfortable with the motion of billiard balls being rigidly determined, and, at the same time, we think we choose to aim the cue stick in a skillful way that depends on how well we line up the shot. If we miss, we blame ourselves. Rationally, we might wonder if perhaps there was actually no chance we could sink the ball that time because a myriad events determined we would miss. When the number of possible causes is too great to consider, we have to ignore them. There is no way to predict whether we will miss again on our next turn. So we keep on playing for the fun of it.

Steven Pinker, a professor in the Department of Psychology at Harvard, thinks the brain is wired to feel like it is making free choices, and he suggests that we should listen to what our brains are telling us. "The experience of choosing is not a fiction, regardless of how the brain works. It is a real neural process, with the obvious function of selecting behavior according to its foreseeable consequences," he explains (1997). It does not matter whether or not our actions are all determined by past events; we have to act as if we had free will. Moreover, the laws and regulations of society are based on the assumption that each individual is responsible for his or her actions. The rules take free will for granted, and it would be difficult to change them at this stage.

THINKING AND MEMORY

Equipped with a complex brain that generates emotions, feelings, memories, and a consciousness of self, humans are the best thinking machines on earth. Thinking opens the doors to the other side, to fantasy and creativity. We can imagine ourselves hunting a deer, lying on a beach, or reading a book. We can play out a fictitious movie in our minds and see how we predict the characters will respond. When we are actually in the woods and a deer appears, we can act out the role of the hunter that we imagined and shoot a straight arrow. William James argued that intelligence results from the use of a large collection of evolved neural circuits, each adapted to a specific problem. Face recognition, discussed previously, is one example. Modules could be added piecemeal in response to new challenges, and new ones could be generated by experiences during a lifetime (Koch 2003). Humans acquired superior intelligence by adding more such modules and perfecting combinatorial use of multiple modules to generate flexible solutions to more varied problems. Progressive addition of modules has been compared to adding subroutines to established computer programs to allow a greater range of freedom in their applications. It seems to make sense but requires rapid generation of a multitude of specific neural circuits that can all function unconsciously. It has been only 6 million years since chimpanzees and humans split from a common ancestor, and yet there is a huge difference in their apparent intelligence. Chimps can solve simple puzzles about as well as a four-year-old child, but they cannot compete with a smart adolescent, let alone a rocket scientist. Where did all those new modules come from in such a short time, and how are they wired?

Although intelligence is notoriously hard to measure, it is likely that a modern educated adult would be considered smarter than a hunter-gatherer living fifty-thousand years ago. The average person these days can tell time accurately, use and invent complex tools, read, and write. Could a cave man learn these skills if given the proper education? If good students simply use a large number of preexisting modules for new purposes, doesn't this relegate modules to general purpose brain functions? It seems to me that the concept of wiring together a critical mass of specialized modules does not explain the rise in intelligence any better than an increase in general connectivity of an expanded number of neurons in the prefrontal cortex. Comparisons of the human brain to computers may fascinate computer scientists, but there is a qualitative difference between the intelligence of a computer program and the intelligence of a computer programmer. I think we have to look elsewhere to find the basis for human intelligence.

The chimp genome is about 99 percent identical to that of humans, but the 1 percent difference allows for thirty thousand changes, any one of which might be critical to the species. Ajit Varki, a professor at the University of California, San Diego, and his colleagues found a candidate gene that had suffered a mutation specifically in humans (Gagneux et al. 2003; Varki and Altheide 2005). The gene is necessary for the addition of a sugar compound to proteins on cell surfaces, and so it could have far-reaching consequences. The deletion appears to have occurred about 2 million years ago uniquely in the human lineage after divergence from the great apes (Chou et al. 1998). Chimps, gorillas, and orangutans all have the enzyme, and they modify their surface proteins by addition of the specific sugar group, while humans do not. Shortly after the loss of this gene, the brains of humans started to expand. Since this protein modification has roles in the control of growth and development, reducing it may have been critical for the evolution of the human brain.

Bruce Lahn, who is in the Department of Human Genetics at the University of Chicago, has used experimental and computational techniques to compare rat, mouse, monkey, chimpanzee, and human genes. He has focused on genes that work in the brain in hopes of finding differences that might account for characteristics specific to humans, such as their unusually large brains. His laboratory has found a subset of genes involved in brain development that are evolving more rapidly in primates than in rodents, and the rate of change seems to have sped up in the human line (Gilbert, Dobyns, and Lahn 2005). The subset includes genes for neuroreceptors, a gene for developmental signaling, and two genes known to be involved in brain growth. One of the genes for brain size regulators, microcephalin, carries a genetic variant seen only in humans. The specific mutation appears to have arisen in an individual thirty-seven thousand years ago and then spread throughout the species. The other brain size gene had a small change in its sequence about six thousand years ago that has been swept to high frequency all over the world (Mekel-Bobrov et al. 2005). Most changes in genes never become prevalent, so this one is exceptional. There appears to have been a strong selective advantage to individuals carrying this modified gene, perhaps because they had bigger and better brains. Lahn thinks that the human brain is still undergoing adaptive evolution. I wonder how we will think in ten thousand years.

With our bigger brains we can process more information, learn more, and remember more. Memories are built when integrated sensory and emotional inputs activate neural circuits in the hippocampus, which acts as a clearinghouse for dispersal of memories to the rest of the brain for storage. Only a small proportion of the total neural activity results in memories. Nerves fire, nerves respond, but no stable trace remains. However, if a nerve repeatedly excites another nerve, there is a chance that the connection between the two will be strengthened. Donald Hebb, the influential Canadian psychologist, made a cogent argument in his 1949 book, *The Organization of Behaviour*, that persistent excitation of a neuron might result in increased efficiency of the connection with the exciting neuron and any other neurons that connect to it. Hebb proposed that the number and arrangement of molecules at the synapses where nerves connect changed as the result of excitation and provided a sort of memory of the event. Subsequent studies showed that long-term potentiation in the hippocampus depends on a persistent train of excitatory impulses or near-simultaneous impulses from multiple neurons.

Biochemical studies have indicated that some of the synaptic changes result from activation of key enzymes, including protein kinases. The initial biochemical changes are reversible, and so the slate can be wiped clean after some time. Hebbian learning suggests that new information is coded by changes in local synaptic strengths. It can account for learning by association, in which two stimuli, say food and the sound of a bell, are presented simultaneously, as Pavlov famously did with dogs. The dogs soon learned to expect food when they heard the bell, and started to salivate. The first step in this conditioned response may have been Hebbian learning in the hippocampus. From there the information is dispersed over large areas of the brain for long-term memory.

I remember an afternoon more than twenty years ago in Paris when I had a *croque-monsieur* for lunch and was fairly violently ill about a half hour later. For years I could not bear to eat a *croque-monsieur*. This sounds like sensible aversion learning; but the trouble is, I knew that my nausea had nothing to do with the *croque-monsieur*. I had often had this type of grilled ham-and-cheese sandwich for lunch with no ill effects. Moreover, one of my daughters had the stomach flu that fateful day, and my wife came down with it the next day. It was clearly just a coincidence, but my brain had registered that a *croque-monsieur* made me sick—avoid *croque-monsieur*! And the memory persists to this day, although I have overcome my aversion to this particular type of sandwich. What started as associative learning, perhaps with

Hebbian connections being established, ended up in memories that surface every time that I see a *croque-monsieur* on the menu.

A lot of what goes on in our brains is unconscious. There are fully formed images and neural patterns acquired through experience to which we do not attend. They may be remolded and reworked without ever becoming explicit. But they are stored in memory and can affect our thought patterns as much as inherited instincts. We are just not aware of them.

Studies with patients in whom large parts of the hippocampus have been lost, as in Damasio's patient David, have shown that establishment of long-term memories depends on neural circuits in the hippocampus. Studies with other patients in whom cortical regions have been destroyed by disease or injury have shown that some types of memories, such as the names of specific people or things, can be lost without affecting other memories. These results should not be interpreted as indicating that proper names are stored in one particular place in the brain and the names for fruits and vegetables are stored in another, only that retrieval of them follows a route through those parts.

Memory has been considered to be a network of stored representations recalled by cueing the memory system. Recent fMRI studies on subjects asked to memorize faces, places, and things have shown that somewhat different parts of the brain are active for each category. When a subject was asked to recall a specific face, the parts of the brain used to store information concerning faces became active. The same was true for places and things, but slightly different parts of the brain lit up. The reappearance of a given category's activity pattern occurred several seconds before the memory was verbally recalled. It seems that category-specific cues are used to retrieve patterns from memory. Subjectively, a choice of memories is presented that the brain rapidly sorts out to find the one which fits the question.

The majority may be rejected within a split second, leaving only a few to chose between.

Recent neuroimaging shows that brain volume and cognitive performance is well maintained until at least sixty years of age. Thereafter, there is a gradual decline in both brain volume and function, but the individual is not significantly impaired for many years. Healthy brain aging may result from its initial large size reducing the likelihood of developing dementia. We start out with more brain than we need (Allen, Bruss, and Damasio 2005). Nevertheless, as we approach old age, our memory fades. Somewhat surprisingly, older people have more difficulty remembering recent events than events that happened long ago, some as far back as childhood. A common form of senile dementia is Alzheimer's disease, which now affects more than 50 percent of people over the age of eighty-five. Everyone forgets things for a while but remembers them later. Those with Alzheimer's never remember them. They will ask the same question over and over, each time forgetting that they just got the answer. They forget simple words and get lost on their own street. As the disease takes its inevitable course, they forget what numbers signify, what goes in the sugar bowl, and become irritable, suspicious, and fearful. "Who are you?" they will shout as a loved one comes in the room. Finally, they forget who they are and their own sense of self dissolves. At this stage they need full-time help to be fed, clothed, and given comfort. The terminal phase may go on for years, with the patient requiring more and more help every day. Dignity is lost, but the body goes on.

EUTHANASIA

Modern medical techniques such as respirators and intravenous feeding have allowed patients to continue to live who would previously have died rather quickly. Families and their doctors are now faced with deciding whether to continue extraordinary efforts to maintain life or to terminate the treatment and allow the patient to die with some dignity and not linger uselessly. If the hopelessly ill person had previously left a record indicating that he or she did not want to be artificially maintained after it was clear that recovery to the point of a meaningful life was impossible, the decision is a little easier to make. However, most people prefer not to think about it and leave no instructions. Of course, if the person can express her own wishes to go on living as long as possible, that's it.

Some years ago, an elderly woman that I loved suffered a massive stroke and was taken unconscious to the hospital. She lay in a fetal position as her family stood around and bid her good-bye. An electroencephalogram had registered no brain activity. Her flatline permitted the attending doctors to declare her brain dead and allowed them to recommend that the respirator be turned off. There was no sustainable life and no hope of any recovery. The family consulted in the room next door, and all agreed to let death occur naturally. The respirator was turned off, and she died almost immediately. It was sad and painful to have to make the decision, even though there was no question that it was right.

Most cases are not so simple. Strokes can put a person in a coma where consciousness is completely absent but life goes on. Now and then the patient may display some reflex motions, giving hope to those around, but false hope. Sometimes, after a few weeks, stroke patients emerge and regain their basic bodily functions. A few improve over time until they have regained many of their physical and mental functions. Others enter a permanent vegetative state with loss of all cognitive ability, but still retain the ability to breathe and make spontaneous movements. Those in a permanent vegetative state show no brain activity. They are living nonpersons. After they are maintained in this state by continuous intravenous feeding for months, it is time to consider terminating the treatment if the patient had previously made it clear that she did not want to be maintained when all hope of recovery was gone (Gazzaniga 2005). Allowing a person to die in such cases is accepted practice in most societies, but it still results in vocal protests from a minority.

A patient with intolerable and lingering pain, and no way out, may request assistance in dying. Voluntary euthanasia in such cases is presently approved in Belgium, Holland, and Switzerland. In the first seven years after the state of Oregon enacted the Death with Dignity Act, there were 208 reported cases of physician-assisted suicide. In other parts of the world, physician-assisted suicide and euthanasia have been practiced for generations without much discussion. The ancient Greeks and Romans accepted suicide of the terminally ill and permitted physicians to help when needed. Throughout the Middle Ages and the Renaissance, it was common practice to let nature take its course and not interfere. At times, strong medicines were administrated that hastened death. Only recently has there been opposition to assisted dying on the grounds that euthanasia of the terminally ill might be driven by greed rather than love. Those who stand to inherit from the patient might encourage premature death. Physicians anxious for organ transplants might assist so as to be able to harvest needed parts. However, there are strict guidelines for voluntary euthanasia that must be followed, where it is allowed. They include cases where a competent patient makes a voluntary and informed decision to die, the patient's suffering is unbearable and there is no acceptable way to alleviate it, and the physician's prognosis of a terminal illness is confirmed by another physician. The guidelines suggest that all three conditions should be met. Even with these guidelines, it has not been possible to pass legislation in many countries that allows for a merciful death.

The problem is compounded by patients with serious mental or physical problems who want to die because living with the constraints and medication is intolerable. They may not be in pain or have a terminal illness, but confinement to a bed or wheelchair robs them of any meaning in life. Their condition may have reduced them to undignified dependence on others for even the little things in life. They would prefer to be dead. A quadriplegic who cannot move anything below his neck may tire of watching television all day and listening to the constant rhythm of the respirator. Continuous nausea, incontinence, and drowsiness induced by medication may become intolerable. A person who has continuous frightening hallucinations and has to rely on drugs that remove all ability to think and respond may chose death when off the drugs. Should these poor souls be given help in dying when they ask for it? The trouble is that many people go through transitory periods of deep depression, and we want to help them recover. Euthanasia is final, irreversible. No mistakes should be made, and the line is not clear between instances where the wishes of the patient should be followed, and where they should be ignored.

There are also cases where forced euthanasia has been practiced. Totalitarian regimes, including Nazi Germany, once routinely killed severely crippled or retarded people. Those with birth defects that would lead to early death, those without functioning brains, those so mangled by accidents that there was no hope of recovery were euthanized in the best interests of society. However, the slippery slope soon led the regime to include presumed deviants and minorities in efforts to "purify" the race. Eugenic arguments were used to justify the practice, and it was dangerous to object. This sorry episode in history has tainted all subsequent decisions concerning the right to life. Rigid brakes have to be kept on euthanasia to avoid a descent in barbarism. However, when due consultation with the patient, family, proxy, and physician results in a clear case for mercy killing, there should be processes in place to carry it out.

When consciousness is lost because of brain injury, is life still there? Do artificially assisted bodily functions constitute the person? The self is absent in patients in a persistent vegetative state, and there is no sadness, no happiness, no nothing in the body. Why should it be maintained at great expense and suffering to the surviving family and the society in general? Some might hope for a miracle, but miracles don't happen very often. An incurable disease or injury that precludes any possibility of regaining any semblance of the previous quality of life can be recognized and the wishes of the patient followed.

For thousands of years the medical profession has followed the Hippocratic oath: "I will prescribe regimen for the good of my patients according to my ability and my judgment and never do harm to anyone. To please no one will I prescribe a deadly drug nor give advice which may cause his death." But times have changed with the invention of sophisticated machines that can draw out the dying phase of life and medications that can stave off infections in a living nonperson long after the self has been extinguished. Decisions based on understanding what is life and what is consciousness can lead to new rules to guide the patient and the physician alike.

Acting as if we had free will puts the burden of decision making directly on our shoulders. No matter how much compassion we bring to a decision, we are always prone to self-serving motives. Evolution has selected for the reproductive success of the individual, and every action and reaction controlled by instinct is aimed at that goal. We have consciousness and conscience to help guide us, but we should never underestimate our instincts. Recent evolutionary selection in the primate line that has led to *Homo sapiens* appears to have favored cooperativity in certain situations, and humans have learned to reinforce these positive attributes to allow us to live harmoniously in large groups. Life is a continual conflict between maximizing short-term benefits to the individual and long-term benefits to the group. They lead into the even more controversial problems of absolute and relative morality that color everyday existence in every society. These matters are the subject of the next chapter.

SEVEN

SELFISHNESS AND COOPERATIVITY

MORALITY IS AN IMPORTANT PART OF LIFE, SOME SAY THE MOST important part. As social beings we need principles to function together effectively. Morality and ethics are learned cultural matters but can be considered in terms of the underlying biology. In high school, I read the great books that raised moral problems and presented a range of solutions. After class and often late into the night, I discussed them with friends and, like most adolescents, came up with jokes, witty pronouncements, and some serious summary statements. Most of these were inconsequential, but one conclusion has stuck with me ever since. It seemed to me then, and it seems to me now, that there is no such thing as a completely unselfish act. Assuming we have free will, then every action we take is because we choose it. Helping an old lady across the street and tutoring disadvantaged children in the inner city are not unselfish acts. We do them because

FIGURE 7.0 *St. Francis Receiving the Stigmata*, by El Greco, 1590. Photograph courtesy of the National Gallery of Ireland.

they make us feel better. Sharing half my dessert with a friend is tinged with the hope that she will share hers with me at a later time. The only act that I could think of as unselfish was an instinctive or conditioned response, such as catching the arm of a stranger who tripped, and then walking on and completely forgetting about it. If I remembered it later and felt I had done a good deed, then my satisfaction with being unselfish would make it into a selfish act ex post facto. So if I have ever done something unselfishly, I can't remember.

On the other hand, there are certainly selfish acts. I have done things for my own good that were detrimental to others. I have avoided unrewarding jobs that someone had to do. I have gone to the lab and left the dirty dishes for others to wash. I have stayed late writing in my office when others wished I was home. I am sure that other people can think of many other cases where I acted selfishly. It is always easier to think of the selfishness of others. You remember the selfish person who pushed ahead in line, but tend to forget the time you inadvertently joined the front of the line yourself.

A person who fails to cooperate and appears to act in self-interest is called selfish. On the other hand, a person who cooperates is not necessarily acting altruistically; it may be directly to her advantage to cooperate, and so there is no cost. Pure altruism is rare or nonexistent in the animal world (Hammerstein 2003; Gazzaniga 2005). A prairie dog that stands up and whistles in alarm when a hawk flies over is obviously putting itself in danger while helping surrounding prairie dogs dive for cover. However, most of the prairie dogs within earshot are blood relatives of the sentinel. Their survival increases the chances of the common genome being propagated and the family surviving. This is kin selection. It can lead to genes resulting in instinctive altruism, but only altruism directed to close kin.

Reciprocal altruism can occur between unrelated individuals when there is an expectation of the favor being returned in the future. The cost to the individual acting altruistically is offset by the return benefit. This requires trust in the fairness of the others, which can easily be broken by cheaters who accept the favor and leave. A society must have a way to punish cheaters to maintain reciprocal altruism. Without sanctions, even those who prefer cooperation may tire of supporting the freeloaders and start to cheat. It usually takes multiple encounters to recognize cheaters and ostracize them.

SOCIOLOGICAL GAMES

Considerable efforts have been made to understand the psychology of selfishness and cooperation in order to predict the outcome of an interaction. There is a game called Prisoner's Dilemma that was constructed at the Rand Corporation to develop strategies for cold war interactions in the 1950s. Scientists working at this quintessential component of the military-industrial complex came up with a devilishly subtle set of rules to see how politicians and normal citizens would react.

In the Prisoner's Dilemma, two robbers are caught in a bank and taken to separate prison cells. They had previously agreed to cooperate with each other and not rat on the other. The prosecutor wants each one to plead guilty and finger the other to cut down on court time. He promises each prisoner that if he pleads guilty he will get a lighter sentence. He wants each one to defect from his prior agreement individually. In fact he promises each prisoner that, if he confesses and fingers his partner, he will be let out immediately if his partner pleads innocent, because his testimony will have closed the case. The case is so weak that, if they both plead innocent (cooperate with each other), they each get

Partner (P)

	cooperate	defect
cooperate	1 for You 1 for P	3 for You 0 for P
defect	0 for You 3 for P	2 for You 2 for P

You

Prisoner's Dilemma

FIGURE 7.1 Prisoner's Dilemma. This sociological game is for two players. Possible outcomes for the game are given in jail time in the four boxes.

one year in jail. If they both confess (defect from their prior agreement), they both get two years in jail. If one confesses and rats on his partner while the partner cooperates with the prior agreement and pleads innocent, then the one who confessed gets off scot-free while his partner gets three years in jail for robbery and lying about it.

If you do the arithmetic, the logical thing is to plead guilty (defect). That way you either get no jail time or two years, depending on what your partner says. But if you plead innocent (cooperate with your partner), you get either one year or three years, depending on what your partner says. On average, it is riskier to plead innocent (cooperate). You can try to guess what your partner will do and conclude it is likely that he is smart enough to defect and so you will both be sentenced to two years. On the other hand, if he is counting on you to cooperate and pleads innocent, it is still better to defect because you get off the hook. If both plead innocent, both go to jail for a year. But you have to rely on the cooperation of your partner to benefit from cooperating. So you defect even if you morally favor cooperation.

There are many variations of Prisoner's Dilemma that have interesting outcomes. One of the more informative is to play the game many times over with each prisoner trying to keep his years in jail as low as possible. When the players are told to play twenty rounds, they usually start off cooperating and defect only toward the end. The fMRI scans taken during the game showed that brain areas associated with reward processing were activated during mutual cooperation, suggesting that the neural network positively reinforces reciprocal altruism (Rilling et al. 2002).

When the prisoners are not told how many times they will have to play, they usually start off by cooperating and continue to do so as long as the other does so as well. But when the partner defects, then they defect on the next round to punish their partner and encourage cooperation. They continue following the other's lead until the end of the game, cooperating only now and then when the other has defected. Computer simulations of this variation of the game have shown that this "generous tit-for-tat" strategy keeps the total jail time to a minimum. It has an uncanny resemblance to the Old Testament "eye for an eye" approach to punishing and deterring noncooperative behavior.

In larger groups many people will actually pay to punish those who fail to cooperate. They recognize that such altruistic punishment maximizes benefits to the whole society by reducing the number of cheaters. Experiments have shown that they are usually willing to incur greater costs when the sacrifice is public and appreciated by others around them. Moreover, activity in a specific portion of the brain, the dorsal striatum, indicates that they derive satisfaction from punishing violators (de Quervain et al. 2004; Camerer and Fehr 2006).

It turns out that, as players get to know each other better, cooperation increases. In small communities, where each player knows the others well, a code of conduct develops that encourages cooperation in a manner indistinguishable from rules of morality (Shermer 2004). Those with connections to others cooperate much more often than they defect. Even when strangers are paired, they tend to cooperate if they meet beforehand and perceive the other to have shared cultural traditions. They look for common appearance, common language, and common behavior, and come to expect that they will play fairly by the rules and so can be trusted more than those whom they think of as different. People are innately aware that their moral values are not shared by all. Although each player may be certain that his moral values are correct, he does not trust someone who clearly belongs to another culture. He will not be so willing to gamble on cooperation with someone who has an unknown set of moral values. The deep, universal feeling of cultural relativism clearly comes out at such times.

For centuries, stories have been told that present moral dilemmas. They do not necessarily involve questions of selfishness or altruism but concern questions of internal gut reactions. They do not have answers that are right or wrong, but they must be answered. The following story is one that I find fascinating, because slight variations result in radically different responses (Gazzaniga 2005).

An empty, runaway trolley is barreling down a narrow street with a sharp turn at the end. There are five people at the end of the street, and they will all be killed when the trolley jumps its tracks at the end of the street and slams them into the wall. Halfway down the street a woman is walking along, oblivious to the impending disaster. You are near a switch that can instantly derail the trolley so that it does not get to the end of the street. If you push it, you will be saving the lives of five people. Unfortunately, the derailed trolley will kill the woman halfway down the street. Do you push the switch? Most people decide that sacrificing one life to save five makes sense and push the switch. However, if the dilemma is changed slightly so that there is no switch, no woman halfway down the street, and derailing the trolley can be done only by pushing a person in front of you onto the tracks, most people do not derail the trolley (Shermer 2004). They cannot bear to think of shoving an innocent person to her death even if it means saving the lives of five others. The rules of the dilemma make it clear that you cannot jump on the tracks yourself, because there is a person in front of you. You have to physically push that person to her death to save the people at the end of the street. Very few people think they would be able to do it. This reaction shows how strongly most people feel about violating the rights of others when it comes to physically interacting. A switch is once removed and can be thought of impersonally. The result is the same: one dead, five saved. The death of a person by one's own hand is different from the death of an innocent bystander by a switch. What would it take to get you to push the woman in front of you? What if she is old or a stranger from another culture? Some people say they actually would sacrifice an innocent person, even by shoving her onto the tracks, to save the five. Are these people more rational or more immoral?

We are faced with moral dilemmas every day without even knowing it. In our interacting world, there are often common resources that benefit everyone. However, individual exploitation can deplete the resources so that everyone suffers. Garrett Hardin made this clear in his 1968 article "The Tragedy of the Commons." He invoked a vision of times long ago when each town would have a field of grass near the center where sheep could graze. Individuals were allowed to bring their sheep and fatten them up. Hardworking shepherds would increase their flocks and bring them to the commons. After a while there were more sheep than the commons could support, and it turned into a dirt patch. Everyone suffered. The individual, hardworking shepherds had done nothing wrong, had broken no rules, and had had no intention of destroying the field.

Something very similar to this happened all around the shores of the Mediterranean Sea following the establishment of ancient city states. The stability and cooperativity allowed the population to increase and with it the number of sheep and goats. The heavily wooded grassland all around the central sea appeared to be limitless and bountiful. The shepherds took their flocks to the rich green pastures and let them graze. After a while they cut down trees to increase the pasturelands. The sheep and goats continued to eat and multiply. Soon the landscape took on the appearance we associate with the Mediterranean now—brown, stony hills with isolated trees every now and then. The flocks are gone and the land is ravaged. It wasn't a change in climate that denuded the hills, it was the tragedy of the commons.

Was there a moral defect in the people of the ancient Mediterranean? They meant no harm as they sought to feed their families and bring meat to the marketplace. No one knew that overgrazing would almost irreversibly degrade the land. And even had they known, there was no will strong enough to enforce sustainable use of the common hillsides. It was immorality by ignorance and creeping consequences.

KIN SELECTION

Consider the moral dilemma of the runaway trolley when you are acquainted with the woman halfway down the street or with one of the five people at the end of the street. If the woman on the street who would be killed if you derailed the trolley is your mother, sister, or daughter, there is no way you would push the switch. The five people at the end of the street may die, but you are not going to kill a member of your family. Likewise, if your son is one of those at the end of the street, you will do everything you can to derail the trolley even though it means killing someone halfway down the street. The fact that four others will be saved is almost irrelevant. Protection of kin is an instinct that is deeply embedded in all cultures. It leads to the solidarity of family and tribes and should never be underestimated. Members of a family can expect all their close kin to share with them. If a stranger comes to the door and begs for a little food, it is another matter. Some might share and some might not. It also depends on how much is in the larder—how much we are willing to give up ourselves. The stranger might be a professional beggar looking for a free ride.

Kin selection works even at the simplest level. Consider again my favorite social amoebae, *Dictyostelium discoideum*. Most of the cells make spores that can seed new generations, but about a fifth of the cells sacrifice the chance to have progeny by building a cellulose tube, entering it, and swelling to push it up. These stalk cells become trapped inside as they build walls around themselves to give strength to the stalk. The majority of the cells in these fruiting bodies benefit by climbing up the stalk so that they can disperse when a blowing leaf or low-flying insect happens to hit them. The stalk cells give up the chance to propagate and die. As long as all the cells in a fruiting body have the identical genes, then stalk cells are no more altruistic than the cells in a finger or a liver that will never be able to propagate but which aid in the fertilization of eggs or production of sperm. The genes in the egg are identical to those in the finger, and so the individual adds to the population genetics, although the individual finger cells are fated to die without direct descendents. The leaves on a tree are not altruistic, although only the seeds will give rise to new trees. All the cells in a tree are kin; in fact, they are a genetically identical clone.

However, some *Dictyostelium* fruiting bodies contain cells with different genetic histories. One strain might have been selected for generations to survive salty conditions and may have acquired genes for resistance to salt. Others might have specialized in eating an unusual type of bacteria. There will be a certain degree of genetic heterogeneity in the aggregates of previously independent individual cells. What is to keep genetic variants from being selected that refuse to go into the stalk and swell up? They would not have to sacrifice themselves for the benefit of cells with a different genetic history, and they would have a greater chance of spreading in the environment. But if these cheaters went off and founded new populations, their progeny might not be able to form a stalk at all, and the spores would just lie in the soil and not spread around. Strains of cheaters have been found that ignore signals to become stalk cells when they develop together with cells of a different genetic makeup, but they form normal fruiting bodies when they find themselves alone and make genetically pure fruiting bodies. The trouble is, the spores they make are defective and do not survive long in the wild (Foster et al. 2004). It looks as if genetic variations that make for perfect cheaters are so rare that they seldom arise. Natural selection has developed strong barriers to cheaters. These soil amoebae have evolved the ability to recognize most potential cheaters and kick them out of the aggregates. It is a form of innate immunity that works much like our rejections of skin grafts from anyone other than our close kin.

Similar mechanisms protect sponges from being taken over by more aggressive sponges that happen to live nearby. Many sponges are colonies made up of multiple individuals, but they strongly resist being joined by others they recognize as different. This is kin selection at its simplest molecular level.

Many social insects generate sterile workers that devote their lives to caring for the queen, constructing and protecting the nest, foraging for food, and tending larvae, while never having larvae themselves. These altruistic workers do not behave in ways that increase their own chances of reproduction, only that of the queen. However, the queen is their mother or sibling, and so the genes she passes on are the same as theirs. Cooperation benefits the group as a whole but can be put in danger by selfish cheaters who get a free-ride on the altruism of others without making much of an effort themselves. When their number gets too high, the group will suffer and produce fewer offspring. From the perspective of evolutionary genetics, the only kind of cheating that works in their world is when a strange queen comes to reside and then is serviced by all the workers in the hive. The defense against such usurpation is based on recognition of odors and shapes that distinguish kin from non-kin. It works most of the time.

Like the apparently altruistic prairie dogs who sound the alarm when a hawk flies over, vervet monkeys also sound an alarm but have a wider range of calls. They give a loud barking call for leopards, a short cough for eagles, and a chutter for snakes. Each of these noises can alert the predator to the whereabouts of the sentinel and put it in danger, but the alarms may save the sentinel's family from being eaten. When the monkeys hear a chutter, they stand up on their hind legs and look in the grass for a snake. When they hear a bark, they run to the trees to escape a leopard. When they hear a cough, they look up to see if an eagle is swooping on them. Since vervet monkey almost always forage together in family groups, the dangerous but effective warning signals save the lives of kin. It is clear how such instincts gave a selective advantage over millions of years.

Humans appear to have inherited many of the same instincts during the long period before the rise of civilizations. The mother's instinct to nurse a newborn baby is universal and benefits both mother and child. The baby's suckling induces the release of oxytocin, the "love and cuddle hormone." This small peptide hormone promotes maternal behavior, induces the flow of milk and gives a general feeling of well-being that leads to bonding with the child. Oxytocin also increases in the suckling baby, reinforcing attachment to the odor of the breast and the sounds and touch of her mother. High oxytocin leads the brain to develop in such a manner that for the rest of her life she will be better prepared to handle stress. Fathers also get an oxytocin high from handling the child. The levels of other hormones, such as vasopressin and prolactin, also go up in participating fathers, which leads to caregiving behaviors and a reduction in libido. While nursing itself delays the return of ovulation, reduced libido plays a major role in better spacing between children. A lot of love is hormonally controlled.

Throughout the world, the basic social unit has always been the family. Brothers and sisters will naturally defend each other and try to protect their parents. Fathers and mothers will sacrifice themselves for their children. When relatives are invited for dinner, their faults are overlooked and they are included in all the jokes and schemes of the family. The extent of the concept of family varies from one culture to another, depending on the traditions and the environmental challenges. In some cases it does not extend past first cousins, while in others it may include all those who trace their lineage to a patriarch who lived long ago. Generally speaking, the more distant the relative, the less cooperation is given as a matter of course.

There are present-day governments based almost exclusively on kinship. A dominant tyrant or sultan may have had hundreds of sons, and their sons and grandsons may fill all the government posts. The assumption is that you can trust your cousin and your cousin will trust you. It is not clear how many generations such a system may survive. Although people in many nations are likely to share a common ancestor with most other citizens, they seldom trace their lineage back beyond their four grandparents. If they went back ten generations, they might see that at least one of their thousand ancestors was also the ancestor of their neighbor. Go back three thousand years and the hundred generations would constitute a web of kinship that encompassed millions. It would be nice if we thought of kin selection in such terms. However, cheaters are a continuous challenge, since the black sheep in the family will try to maximize their own gain. They have to be recognized and punished to deter further selfish acts. There appear to be instincts for punishing or eliminating those who are a threat to cooperation.

Consider again the dilemma of the runaway trolley. Let's say you know that one of the five at the end of the street is a convicted serial killer. You may consider letting the trolley go on and smash them all, even if four innocent bystanders die. If all five are convicted serial killers, you almost certainly will let the trolley crush them. What about the dilemma of having four convicted serial killers and a close family member at the end of the street? The instinct for kin protection will likely win, and you will push the switch to derail the trolley.

Emotions as well as rational thought can drive decisions. This has been shown in a statistically significant way by people involved in the Ultimatum Game. In this game, one player is given a hundred dollars and told to split it with another player, a stranger, who is in another room. The first player, the proposer, can decide to split it fifty-fifty or ninety-ten or at any other rate. The proposal is transmitted to the other player, the responder, who then decides to accept it or reject it. If the offer is rejected, neither of them gets any money. The rational decision for the responder is to accept whatever is offered, because even a little is better than nothing. However, when less than 10 percent is offered, most responders feel sufficiently insulted that they choose to punish the proposer by rejecting the offer so that the proposer also gets nothing. The wish to punish someone who makes a selfish offer appears to depend on functions that occur in the dorsolateral prefrontal cortex of the brain. When this region is inhibited by a focused magnetic field, the responder can still recognize when an offer is unfair but is less likely to reject it. Functions in this part of the brain appear to be necessary to override rational decision making and promote fairness.

Although almost all offers of 40 percent or more are accepted, offers of 20 percent are accepted only about half the time. However, if the responder is allowed to insert a note, such as "Not fair! You're selfish!" along with the answer concerning acceptance or rejection, then the acceptance rate for an eighty-twenty split goes up to 80 percent (Xiao and Houser 2005). It seems that we want to punish those we perceive as unfair, and if we have no other way of showing it, we will deny them any benefit, even if we lose out as well. This is costly punishment. But if we can send a message of disapproval to the proposer, it can act as a satisfying alternative form of punishment. Moreover, expressing anger at the low offer keeps the proposer from interpreting the acceptance as a sign of inferiority.

Bitching and moaning is a common occurrence in daily life, even if it does little to change the situation; it allows us to let off steam and avoid frustration. We want to curse those who act unfairly, partly to punish them and partly to calm ourselves down. We also enjoy expressing positive emotions when we find that others have treated us fairly. Society always seems to run more smoothly if healthy emotions are freely expressed.

EVOLUTION OF COOPERATIVITY

Darwinian evolution has emphasized the role of competition among individuals. However, cooperation is one of the most important determinants of human success. While selfish acts are expected in a capitalist world—think of the CEO's mantra "Maximize profits"—we could not do business if we didn't trust each other most of the time. Although billionaires are certainly selfish, they make their money through cooperation. Even the exchange of money is an act of trust and cooperation. The paper that money is printed on is hardly worth anything, but a billion dollars can buy you a lot of things.

Selection favors those who survive and reproduce; others are eliminated. A dominant lion will take the choice piece of meat from a fresh kill before letting weaker members of the pride have a bite. Males will fight to the death for access to the females. However, a lioness will bring back meat to her cubs and defend them at peril to herself. Cooperation for the benefit of others is seldom seen except among close family members, where kin selection clearly benefits survival of the family genes. Reciprocal altruism in nonhuman animals is restricted to brief periods among small numbers of individuals. Even lions will hunt together in groups of a half dozen when the mutual reward is more wildebeest kills. If the group is too large, individuals cannot keep track of reciprocity and so do not show reciprocal altruism.

Humans have developed elaborate ways to encourage cooperative behavior. We call it good. And we discourage noncooperative behavior by calling it bad. A good person follows the norm and leads you to believe he always will. Past experience shows you that you can trust a good person. When you do a favor for a good person, you feel you will be suitably rewarded in the future, one way or another. A bad person breaks the rules repeatedly and does not return favors. Such a person will get a bad reputation and seldom be invited to join in communal ventures. Group cooperativity depends on recognizing and punishing those who fail to cooperate. Otherwise, the number of cheaters will increase and previously good members may fall into bad habits.

A person is perceived as good if she cooperates and bad if she defects. But the person might not feel she was good for having cooperated; she might feel she was just a dupe for being coerced. Likewise, she might not feel she was bad for acting selfishly; she might feel it was her right, or that it was important for an overriding reason such as self-preservation. Good and bad are not always clear-cut. There is a gradation from very good to very bad, but we would probably only agree upon the extremes. When groups, such as states or religions, make the rules, they set the scale. Murder is worse than stealing and stealing is worse than lying. However, the laws are often made for the benefit of the group and not the individual, and the individual may suffer in some cases. Since the time of Hammurabi in Babylon, societies have attempted to make the punishment fit the crime, but the laws still vary widely in different places.

Emotions that evolved to modulate simple behaviors in primates have been reinforced in humans to maximize cooperation (Fessler and Haley 2003). Darwin recognized that anger, disgust, fear, surprise, and happiness are expressed in much the same way by chimpanzees and humans. These are inherited responses to the world around us. Facial expressions are universal giveaways. Few of us are poker-faced when making significant decisions, although actors and con artists have learned to fake emotions at will. Without the social emotions guilt, shame, empathy, and remorse, we would all be sociopaths, and human society would not exist (Bowles and Gintis 2003).

Anger at defectors can motivate punishment even if it is costly. We angrily curse someone who harmed us in hopes that humiliation will deter repeat offenses. We may be so angry that we try to ruin his reputation even if it costs us credibility. Disgust is closely related to anger, although it evolved to avoid contamination in foods. It has been extended in humans to include the reactions to unacceptable behavior such as torture, child molesta-tion, and exhibitionism. Expressions of disgust or contempt are particularly shaming and can act as a form of punishment. These emotions are raised almost unconsciously by behavior that is beyond the pale of accepted practices. Their expression can act as a strong deterrent to those who have failed to conform to societal values.

Expressions of happiness can reinforce cooperativity and solidify trust. Laughing at the same jokes, cheering the same team to victory, and rejoicing in the birth of a child can consolidate cooperation. Pride is an emotion that motivates compliance with norms. It can be a completely internal feeling or require an audi-ence to be rewarding. The feeling you get when awarded a prize reinforces your willingness to follow the rules of the group that made the award. Standing with others praising a shared symbol works to bring many people together. We are asked to love one another and told that, if we do, we will be rewarded by a feeling of righteousness. However, if pride and righteousness get out of hand, they can lead to intolerance and the bullying of others. Moral outrage is a similar feeling, one that can be effective in deterring defectors but can quickly alienate others. Envy can dis-rupt cooperation by leading to unrealistic hopes of reward. These universal emotions underlie many of the interactions of men and women in all walks of life. Societies have encouraged some and discouraged others, to allow their members to work together for the common good. Rituals and ceremonies have grown up to reinforce the group values and ostracize the selfish.

Cooperation has to overcome the impulse to defect for short-term gains, behavior that weakens mutual trust. Humans can think ahead much better than other apes, and so can temper their urge to defect as they see the potential consequences of their actions, especially if they think they might be caught. In general, humans are more docile, polite, and cooperative than other great apes. Chimpanzees, even those raised in human households by kindly surrogate parents, do not learn to wait their turn or share with others. They take what they want and will even bite the hand that feeds them. Humans, on the other hand, are good at imitating the behaviors of respected elders, recognize accepted practices, and attempt to follow them. Such traits allow cooperative behaviors to be culturally inherited.

In the last million years, multiple hominid species arose in Africa, but few were sufficiently successful that they multiplied and spread out over the world. It may have been the mix of emotions and innate behaviors of *Homo sapiens* that allowed a high degree of cooperation within the tribes so that they could displace other hominids and survive. When Cro-Magnon tribes left Africa about fifty-thousand years ago, they carried the genetic predisposition to function together over extended periods, which allowed them to survive in the rapidly changing climate of the Ice Age. Big game hunting required close, dependable interactions. Raiding and warfare even more so. Tribes with a high proportion of individuals who looked after only themselves were defeated and their members assimilated into more cooperative tribes, either as wives or slaves. When the agricultural revolution allowed populations to increase dramatically about ten thousand years ago, dominant tribes were able to construct the institutions that reinforced basic instincts for cooperation. Family lands were tended for the benefit of future generations and not stripped bare. It was foresight, not morality, that led to good husbandry.

The rise of nations depended on the degree to which individuals recognized the advantages of membership and avoided sanctions by following the norms and practices. Fortified towns and cities required watchmen and rapid responses to calls to arms. Social institutions broke down at times, resulting in civil war, anarchy, and dissolution of the state. New ones arose on the ruins. As technology improved and travel spread out farther, cities banded together for trade and protection. The borders of such nations changed continuously, but a sense of patriotism held the populace together in the face of outside threats. Symbols such as flags and songs were invented to signify membership, and the styles of dress and facial hair took on new meaning as they came to be used to recognize others with a shared cultural heritage. Allegiance with thousands of strangers can be recognized at a single glance over a crowded square or battlefield. Language and books played central roles in defining nations—think of the Bible, the Koran, *The Iliad*, the Bhagavad Gita, and the Tao. Knowledge of a shared history, even if it is partly fabricated for propaganda, can hold millions together.

Nations have used two basic methods to instill conformity: top-down coercive dominance in police states and bottom-up segmental hierarchies in republics (Richardson, Boyd, and Henrich 2003). Both are based on the tribal and kinship values honed over thousands of generations among small groups of people. In police states, decisions of the dictator or oligarchy are transmitted to the captains, who then return to their lieutenants with orders that are carried out by the foot soldiers. At each level of command, decisions are conveyed to a group of no more than a hundred people, where trust and reciprocity can be easily established.

In a good republic, many decisions are made at the village level among a hundred or so neighbors, many of whom are relatives. Problems that involve multiple villages or neighborhoods can be solved by representatives to a citywide or county government. They can send their leaders to a national assembly to make decisions for the whole nation. Those that work best have less than a hundred members. At each level the aspects of human nature that worked at the tribal level can be adapted to larger groups. The danger with this form of government is that representatives may lose contact with and allegiance to their constituents and transfer their loyalty to the governing body in which they work.

Although we benefit from millions of years of evolutionary selection as social animals, we have only rudimentary instincts for cooperativity. We cannot count on others helping us *all* the time. As societies expanded from hunter-gatherer families to city-states with communal irrigation, rules were laid down to regulate acceptable behavior. Unless everyone abided by the rules, the fields would go dry, the granaries would be empty when famine struck, and all would suffer. Each society developed standards appropriate to its needs and environment. In fact, the rules and culture often defined the society. Can we see a common sense of morality that humans have adopted in almost every culture?

"Do unto others as you would have them do unto you" is often called the Golden Rule and is found in various forms throughout the world. It comes in several versions in the books that define different cultures. In Leviticus in the Judeo-Christian Bible, it is said: "Thou shalt not avenge, nor bear any grudge against children of thy people, but thou shalt love thy neighbor as thyself." It is not clear that this applies to all people or only the Hebrew people. Several hundred years later Hillel Ha-Babli wrote, "Whatsoever thou wouldst that men should not do to thee, do not do that to them. This is the whole Law. The rest is only explanation." This sounds more universal. One of the sayings of Confucius is: "What you do not want others to do to you, do not do to others." The Greek philosopher Isocrates put it in a similar manner: "Do not do to others what would anger you if done to you by others." This seems better suited to interacting with others who may not feel exactly as you do. The Christian gospels give a more positive message: "As ye would that men should do to you, do ye also to them likewise" (Luke 6:31).

The trouble with following the Golden Rule stated in this manner is that you might want a man to look you in the eye when greeting you, and so you might look a stranger in the eye upon meeting him. In some cultures this is taken as a sign of disrespect or hostility and may lead to a violent reaction. The Mahabharata, as usual, gives it both ways, "This is the sum of true righteousness: deal with others as thou wouldst thyself be dealt with. Do nothing to thy neighbor which thou wouldst not have him do to thee hereafter." Following both of those rules will probably keep you out of most trouble.

Books teach morality and ethics on a wide variety of cultural matters. They record the rules set down long ago and attempt to make them relevant to modern-day problems. Schoolchildren are drilled in the symbols and values that the society deems to be self-evident. The trouble is that the messages in the books are sometimes self-contradictory and vary widely from one heritage to the next, so that cultures clash when they are forced together.

CIVILIZATIONS

Common languages facilitated the interaction of neighboring tribes. Differences could be discussed and cooperative ventures worked out. Knowledge could be shared and values agreed upon. When agriculture developed on the fertile plains between the Tigris and Euphrates rivers and along the banks of the Nile, it soon became clear that even more fields could be productive if water could be brought to them from the river. Irrigation canals were built and tended by tribes along the shores. Food became plentiful for the first time, and surpluses could be stored for dry seasons. There was little danger of armed raiders coming from the surrounding deserts, since few could survive there. The tragedy of the commons hit when the number of canals and ditches increased to the point that the water level in the river itself started to drop.

Villages near the source could selfishly dig deeper canals, but downstream farmers were out of luck. Only if they cooperated and rationed the water could all thrive. Strong cities set out to enforce compliance with the water rules to maximize the bounty of the soil. Villagers could no longer take whatever amount they wanted, or the river would soon sink into the sands. The temptation to maximize private short-term gain flowed past each farmer who had to wait his turn to open the sluices, and this undoubtedly led to secret watering in the night. A force that could police the irrigation ditches day and night was needed, and it was provided by central leaders who also took responsibility for protecting the storage of grain surpluses until needed in years of famine.

From the beginning of recorded history, five thousand years ago, there have been kings in Egypt and Mesopotamia with armies, police, courts, and courtiers. Dynasties were established as kings passed on their kingdoms to their successors. The farmers did not always give up their surplus grain and other produce gladly, but they paid their taxes to avoid punishment or banishment. They would naturally curse the tax collector, even the king himself, but only when the repression reached a level where the peasants revolted was there a serious threat to the kingdom. Priests were enlisted to bring discipline to the masses and insure the dominance of the lawgivers. Some kings found it convenient to let it be known that they were gods themselves, and that the priests served them. This kept the clergy in line and cut down on the resentment of the masses. The pharaohs took on the mantle of God on earth to cover their human vulnerability and encouraged the idea that they carried a divine soul that extended beyond the physical world into the afterlife. Elaborate ceremonies and triumphal temples were built to reinforce reverence for the pharaohs. As kings came to rely more on priests, religions were established that perpetuated the myths, rituals, and dictates of the state. They enforced the rules with promises of an afterlife or threats of hell. Established religions soon became more permanent than the rulers and took unto themselves the preservation of the faith. Armies were raised to defend one god against others. People were slaughtered in the name of God.

As discussed earlier, humans have the ability to predict further into the future than any other species. Knowledge of past experiences tells them what to expect in years to come. Each year the days get shorter in fall and then slowly lengthen in spring. In the dark days of winter, we know that long summer days will come again. We also know that old people die and that we are slowly getting older. We can predict that one day we too will die. The trouble is, we have a strong instinct to avoid death at all costs. What can we do about aging? We are left with an uneasy feeling of inescapable conflict. Religions have recognized this uneasiness and offered various ways out. They offer an eternal soul to all who believe in them and follow their tenets. Faith in an ethereal entity only weakly tied to the physical world resonates with the feeling of self not tied to the body, and people flock to the temples.

Some organized religions tell their followers that the soul does not end when they die but is reincarnated in a newborn in an endless cycle. Others promise eternal bliss to the righteous. Such a reward is a strong incentive to play by the rules. Religions have also put the fear of God in the minds of their followers by foretelling eternal damnation for those who stray from the straight and narrow. The concepts of heaven and hell satisfied the righteous indignation of those who felt they were unfairly treated by others on earth: they will get their just rewards in heaven, and the selfish will rot in hell. In these ways the church has carried out its role as enforcer of the norms and defender of the faith. Even kings have been kept in line and forced to protect the church to save their souls.

Since the dawn of civilization, suppliants have gone to church with special pleas—for rain, for health, for children, for all the things that were beyond their ability to control. The churches encouraged such devotion and took credit when it happened to rain or the sick became well. One time in Guatemala I visited a shrine in a small village near Lake Atitlán. There was a statue of the Mayan saint Machimon surrounded by burning candles on a table in the small dark house. The statue was dressed in colorful robes with a hat on and a cigarette in its mouth, with a pile of flowers, liquor, tobacco, and money offerings at his feet. Machimon is said to carry messages to the gods when suitable offerings are made. A local Mayan offered a can of beer and some cigarettes as he requested a new carburetor for his car. The old one was broken and he could not afford to buy a new one, explained my hostess, who spoke the native language. The man seemed completely sincere and trusted that the gods would somehow provide for his honest needs. I don't know when or how he got his car working again, but the simple ceremony certainly seemed to ease his anxiety.

Organized religions have come and gone, but religions still play a central role throughout the world in providing guidance to their followers. Only recently have democratic institutions given citizens a voice in the laws and regulations, so that they feel they are masters of their own destinies and accept justice dispensed by the courts. Slowly religions have become less relevant to the functioning of societies. Morality has not suffered as the result of shifting from an absolute, God-given code of conduct to a rational set of shared values. Any existing code of conduct depends on the specifics of the conditions and thus a form of relativism. Many rational thinkers strongly reject moral relativism, and yet they cannot clearly establish a higher authority.

Good and *bad* are words of approval and disapproval. Like a pat on the back or a slap in the face, they communicate the judgment of the moment. As we have seen, continued cooperation depends on recognizing and punishing cheaters. Calling a cheater "bad" is an effective punishment that does not have to be based on any system of absolute good and evil. The defector is considered bad for the cooperative, and that's it. From another point of view, the defector might be considered good. The civil disobedience of Mohandas Gandhi in India and Martin Luther King Jr. in the United States is held up as great moral courage in the face of official discrimination. At the time, those who nonviolently blocked thoroughfares in the United States were beaten and jailed by the police for obstructing traffic and commerce. The Freedom Riders who ignored the directive to move to the back of the bus were considered scofflaws and harassed. Dr. King was sent to the Birmingham jail in 1963. A year later he was awarded the Nobel Peace Prize. For almost fifty years the Dalai Lama has led a nonviolent campaign for the autonomy of Tibet. He was head of state in 1959, when he was forced into exile for not cooperating with the Chinese forces. In 1989 he was awarded the Nobel Peace Prize. A traitor to some may be a hero to others.

For a long time, societies have codified morality in written commandments, constitutions, and laws. Natural laws are derived from the virtues and vices accepted by the society. Many of the Ten Commandments are embedded in the laws of countries around the world. There are laws against murder, stealing, and bearing false witness. If a society determines that adultery should no longer be a capital offense, then the laws are changed. There are laws that ensure certain types of cooperation, such as driving on the right-hand side of the road. Tax laws are enforced to raise the money for public works, including roads, bridges, and police. If all are forced by law to pay, the communal coffers can be filled and pay for a well-ordered society.

Even in a rational society, there is a place for religion. Although Karl Marx considered religion to be the opium of the people, it provides comfort and hope to those facing insurmountable difficulties. In churches, synagogues, and temples, people can find solace from the barbs of life and talk about what matters to them and what troubles them. The music of the choir consoles them and takes them out of their woes. People often need a belief in a supernatural force that will care for them when no one else does. Although they may know that it is irrational to suppose that a deity will intervene in their individual troubles, they pray that it will happen. "Blind faith" does not question the transcendent. The wish for miracles is strong even in the face of overwhelming evidence that miracles rarely, if ever, happen. Stories of religious experiences, including visions of holy figures and voices in the night, abound.

A few years ago my wife and I visited the Mayan ruins of Tikal in Guatemala. We had the good fortune to have Clarence as our guide. He grew up in nearby Belize and had come to Tikal in the late 1950s to work on the excavation of the temples. Although at first a simple laborer, he became so knowledgeable about the architecture and inscriptions on the steles that he was invited to spend several years in Philadelphia at the University of Pennsylvania, where many of the artifacts were cleaned, cataloged, and studied. In 1970 he returned to Tikal and the jungle he loved. He had stories about each temple and the trees, birds, and plants. He showed us a bush that can be used to make hair dye and mushrooms that cure headaches. He was convinced that the power wielded by the priests of Tikal came not only from their ability to predict the movements of planets and stars and the change of seasons but also from their interpretation of signs and portents. He ascribed their magical insights to the effects of hallucinogenic mushrooms and plants. Clarence argued that all religions started with the use of mind-altering drugs. He made a good case for the central role of hallucinogens in early religions, in which the experiences were interpreted as opening the doors to the future and seeing the face of God.

Chemicals in a variety of plants and mushrooms can give access to transcendent, mystical mental states. Inner and outer worlds are experienced as epiphanies and moments of illumination that can lead to enduring changes in emotional well-being and values. The priests soon learned the ingredients and preparation of the holy potions and incorporated them into their ceremonies. Peyote has been venerated for centuries in the Native American Church, where it is used sacramentally. The Huichol Indians consider peyote to be the heart and soul of their creator, Deer-Person. Their shamans ingest peyote and dispense it to their followers as holy medicine. The ritual use of psilocybin mushrooms and mescaline is widespread in Central America, and the hallucinogenic beverage ayahuasca is sacred in some South American churches.

Long ago, the forces of nature were thought to be the whims of gods. Volcanoes, earthquakes, storms, and floods were terrifying and needed to be explained. What were their causes and what were their purposes? To answer these questions, gods were made in man's image. They were held accountable for everything from lightning to the daily passage of the sun across the sky. Gods of all sorts abounded on mountaintops and in the sky and sea. Their exploits formed the stuff of myths and legends, often of the grotesque and fantastic sort. As knowledge about the natural world grew, the old stories took on a quaint irrelevance and science pursued the underlying causes of natural phenomena. We now understand that volcanoes occur at weak spots in the earth's crust where magma rises from the mantle. Temperatures, compositions, and movements in different types of volcanoes have been carefully measured and the forces are fairly well understood. We do not have to blame eruptions on Vulcan any more.

Newtonian mechanics and relativity can explain the formation of the galaxies and our own solar system, but there are questions that even modern physics cannot answer: Why is the speed of light in a vacuum $c = 299{,}792{,}458$ meters per second? Why is

Planck's constant h = 4.135 667 43 × 10^{-15} eV × s? What caused the Big Bang when all mass-energy entered the observable universe? Will the universe keep expanding, or will it start to collapse one day? Answers to these and other questions often invoke God as an all-powerful prime mover acting in a realm far removed in time and space. However, as the great physicist Stephen Hawking once remarked, "We could call order by the name of God, but it would be an impersonal God. There's not much personal about the laws of physics."

Looking for purposes in nature may be the wrong question. It makes no more sense than asking, "What is a mile north of the North Pole?" The answer is outside of the frame of context. The same is true when looking for the purpose of life. It too is the wrong question. Life arose on this planet long ago and slowly changed as the planet around it changed. We do not know the creatures that first swam in the seas, and we do not know with certainty how they evolved into birds, bees, and human beings. We can guess, but there is much that science has not explained. Some would say that it must have been a miracle for bacteria to give rise to amoebae and algae that keep their chromosomes in a nucleus and swim around using flagellae. They ask how complex organs such as eyes could have appeared. They argue that only the intelligence and foresight of God could be responsible for such complex beings. Although science has not been able to replicate in the laboratory all the evolutionary steps, scientists have developed plausible explanations for how small, random changes could have resulted in selectively advantageous structures and functions (Kitcher 2007). Scientists still keep trying to fill the gaps in knowledge, while fundamentalists say that the works of God can be recognized in the gaps, and that we should just accept on faith that miracles happened and stop trying to better understand.

Perhaps the largest gap in our understanding of life is how it began. When Earth was first formed from material orbiting the sun, it was lifeless. Within a few hundred million years, bacteria were present in the mud that slowly turned into rock. Primitive life took over the planet rapidly. We do not know the precise sequence of events that led to the first protocells, but the next chapter presents plausible scenarios for the spontaneous generation of an RNA world that gave rise to life as we know it. Although prebiological science has a long way to go, there is no need to invoke miracles at any step.

Chance mutations over hundreds of millions of years generate the variations that can result in new characteristics and new species. Natural selection insures that only those fit for their lifestyle survive and multiply. As the environment changes, species come and go and give rise to other species. Even the most complicated metabolic pathways, molecular motors, and organ systems can be built up piecemeal without intervention by an intelligent designer. Given sufficient time, mutation and natural selection can gradually lead to the most finely tuned organisms. It was pure chance and selection that gave rise to a species of primates with advanced cognition, language, and the ability to plan, argue, and wonder. Right and wrong have been argued ever since people came together and tried to impose their will on others. But, as a character in John Steinbeck's book *The Grapes of Wrath* said, "Maybe there ain't no sin and there ain't no virtue. There's just what people does. Some things folks do is nice, and some ain't so nice, and that's all any man got a right to say."

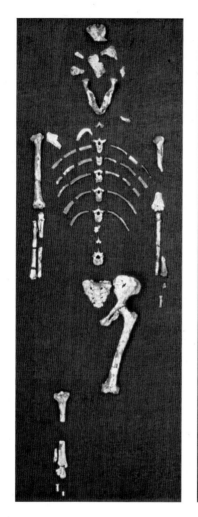 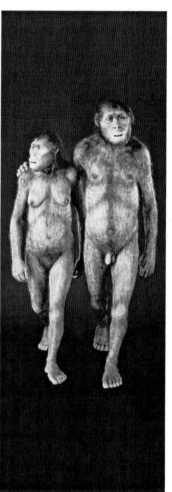

EIGHT

THE ORIGIN OF LIFE
AND EVOLUTION OF MAN

"OMNE VIVUM EX VIVA," WROTE THE GREAT MICROBIOLOGIST LOUIS Pasteur. All life comes from life. In 1859 he presented his ingenious experimental evidence against spontaneous generation of life to the French Academy of Science and won the day. It had previously been thought that rotting meat generated maggots, that mice could be "created" by wrapping dirty rags around wheat, and that all sorts of other strange concoctions would give rise to frogs, lice, and worms. Pasteur showed that, when he

---

FIGURE 8.0 Fossil bones of Lucy (*left*), the species *Australopithecus afarensis*, were recovered from 3.18-million-year-old rocks. A reconstruction of *A. afarensis* (*right*) based on the fossil evidence shows a couple walking upright, side by side. Their footprints were found in 1978 at Laetoli in Tanzania in 3.6-million-year-old rock. Photo at left provided by The Institute of Human Origins, Arizona State University, courtesy of the National Museum of Ethiopia. Photo at right by Denis Finnin and Craig Chesek, American Museum of Natural History, image number 4742(1).

filled a flask with boiled meat broth and then bent the opening into a goose neck, so that it was open to the air but particles could not get in, no microbes grew in the broth. I have seen some of the original flasks in the basement of the Institute Pasteur in Paris, and they are still sterile.

But if all life comes from previous life, where did the first life come from? It is an old question long bound up in religious beliefs. Only in the last 150 years have there been attempts to give a scientific explanation for the origin of life. As long ago as 1871, Charles Darwin suggested that life may have begun in a "warm little pond, with all sorts of ammonia and phosphoric salts, lights, heat, electricity, etc. present, that a protein compound was chemically formed ready to undergo still more complex changes, at the present day such matter would be instantly devoured or absorbed, which would not have been the case before living creatures were formed." Since then, many scientists have tried to apply the laws of thermodynamics and chemistry to the problem, including the biochemists A. I. Oparin and J. B. S. Haldane and the chemists Leslie Orgel and Stanley Miller.

I first met Stanley Miller in 1961, about eight years after he had famously passed an electrical current through a flask containing methane, ammonia, hydrogen, and water and found that, in a few days, significant amounts of amino acids and other building blocks of life accumulated (Miller and Urey 1953). He had demonstrated that complex molecules can be readily generated from gases likely to have been present on the young earth. I was a bit awed as he showed me the actual flask in which these first successful studies of prebiotic chemistry were carried out. In those days we were working in a building at Scripps Institute of Oceanography in La Jolla, California, and we often had lunch by the beach. Stanley would patiently explain the chemistry to me and point out how little was known about the origin of life. Our conversations have continued for the last forty years while we have worked in adjacent buildings at the University of California, San Diego. At times these discussions have included Leslie Orgel, who works up the street at the Salk Institute; Gustaf Arrhenius, who still works down at Scripps; and Jerry Joyce, who works at the Scripps Research Center, just past the Salk, three towering figures in the quest to understand the origin of life. Although no one has been able to spontaneously generate life in the laboratory, ideas and theories based on the principles of physical chemistry and molecular biology have been discussed and tested. The gaps in knowledge of the steps that led to the first living cell are still huge, but this does not deter imaginative, hardworking scientists from trying to come up with plausible steps that could have led from Stanley Miller's primeval soup to a self-replicating entity.

THE RNA WORLD

The first question concerns how long it took for life to appear on this planet, and there are unambiguous answers in the rocks. The oldest rocks that show signs of life are almost 3.5 billion years old. Dating such rocks requires fine ion beams to measure the proportions of elements in zircon crystals that formed in the rocks as they were laid down. Zircons are tough uranium-containing minerals that last through the ages. The uranium[238] isotope decays to lead[206] and has a half-life of 4.5 billion years. So after 3.5 billion years, about a third of it will have become lead. Another uranium isotope, uranium[235], decays to lead[207] and has a half-life of 700 million years, so only 3 percent will remain as uranium after 3.5 billion years. Measuring these isotopes can determine the age of rocks to an accuracy of 1 percent.

The Warrawoona Group of hills in northwestern Australia have rocks that date to 3.464 million years, and some contain microfossils that have been interpreted as ancient bacteria (Knoll 2003). Rocks of a similar age near Johannesburg in South Africa contain bacterial fossils surrounded by organic material likely to have been produced by photosynthesis. These formations also have visible domes of wavy laminations that are similar to the stromatolites that form when successive bacterial mats are covered in shallow seas. It is clear that, soon after the seas and continents appeared, the earth became a biological planet. Stanley Miller once speculated on the time it might have taken for life to originate: "A decade is probably too short, and so is a century. But ten or a hundred thousand years seems okay, and if you can't do it in a million years, you probably can't do it at all," he said. I think he was rushing it a little.

Electrical discharge in the atmosphere is just one way the building blocks of life could have accumulated in seas and ponds. Gunther Wachtershauser has pointed out that pyrite, or fool's gold, can catalyze the conversion of carbon dioxide into organic compounds. Pyrite forms near volcanic vents when hydrogen sulfide reacts with iron monosulfite. In water with dissolved carbon monoxide and carbon dioxide, the organic compounds, including amino acids, form a scum on the surface of growing pyrite crystals. The jump from there to polymers is easy to envision. It has even been shown that small peptides are formed when amino acids are mixed together with iron and nickel sulfides (Huber et al. 2003). Likewise, the nucleoside bases that are the monomer precursors of RNA and DNA are made by the action of sunlight on certain minerals (Senanayake and Idriss 2006). The energy of nature drives simple molecules to combine and recombine, building chemical complexity. The stability of the molecules is as important as their rate of synthesis, since many may have to wait for hundreds of years to polymerize. Nevertheless, the rate of synthesis is critical, because "it takes all the running you can do, to keep in the same place" as the Red Queen told Alice in Lewis Carroll's *Through the Looking-Glass.* This principle applies wherever components must coevolve.

The initial nucleic acid polymers may have used simple three-carbon linkers to string the nucleic acid bases in a chain. In modern-day nucleic acids, the bases are connected by five-carbon sugars, but these linkers are harder to make under prebiotic conditions. The simpler chains could still incorporate the four different bases, adenine (A), guanine (G), cytosine (C), and uracil (U) in specific sequences, so that the polymers had properties different from each other. One might catalyze one reaction, and the next a quite different one. They could each be copied into a complementary strand, as discussed in chapter 1, because A and U preferentially associate, and G and C will form hydrogen bonds with each other. Then the complementary strands could replicate to make a new copy of the original sequence. Copy errors would generate new sequences, which, now and then, might have had useful new properties. There are millions of possible permutations in nucleic acids only twelve bases long, and at some point many of them were made by chance.

Fatty acids also can be synthesized in prebiotic conditions, and they have the tendency to form lipid layers around collections of nucleic acids that are much like cell membranes. They can keep certain molecules in, let others out, and replace them with ones in the surrounding soup. Changes could proceed within each droplet independently of what was going on in nearby droplets. Now and then, droplets undoubtedly got so big that they split in two. When the useful information contained in a droplet was partitioned effectively to two droplets, a hereditary system was not far down the road. The first steps in the origin of life were not really driven by Darwinian evolution; instead the race went to the swiftest. Darwin and earlier naturalists, including Titus Lucretius in 55 B.C. and Jean-Baptiste de Lamarck in 1809, emphasized that competition favored the fittest. However, there was little competition before life arose, only selection for stability and reproduction. The steps could be slow and plodding, since nothing that could eat them had yet evolved. Moreover, there was no oxygen in the atmosphere, and complex molecules could accumulate without danger of being spontaneously oxidized.

Rare droplets in which five-carbon sugars replaced the three-carbon backbone of their nucleic acids would have had a considerable advantage, and when the reactions facilitating such a change arose by chance, they were passed on to more and more progeny. We have no idea exactly how this happened, but it is a very likely step. You can have 10 billion droplets in a liter of seawater, and there are billions of liters on the surface of the planet. Do the math. Even an improbable event becomes probable when the number of possible attempts is astronomical. Give it a little time and it will happen. Droplets that accurately connected the five-carbon ribose backbone of nucleic acids together had RNA. The RNA world was born!

It was not life as we know it, but it probably did very well for a while. The molecular mechanisms for specifying proteins had not yet evolved, and everything was done by the RNA molecules themselves. Amino acids may have been pasted here and there along the RNA chains to expand the number of reactions they could catalyze. Experiments in a variety of labs have shown that RNA can catalyze the synthesis of new RNA molecules and can trim and cut RNA when this is needed. RNA is still used by cells to form peptide bonds, to pick and sort among amino acids, to start the synthesis of DNA molecules, and to speed up all sorts of reactions. No one knows how long the RNA world lasted, but it probably gave rise to some pretty sophisticated droplets.

GENESIS

Peptides are also formed spontaneously under prebiotic conditions when amino acids are strung together in random sequences. Polymerization is facilitated by carbonyl sulfide, a simple volcanic gas that was likely present on the young earth (Leman, Orgel, and Ghadiri 2004). Peptides can be more specific catalysts than RNAs and provide better traffic control of the metabolic pathways. However, since peptides are not able to make copies of themselves, even those that helped a droplet to manage its affairs would be diluted out as the droplet grew and divided. They might appear again by chance, but not in a reliable manner. RNAs can replicate but need a mechanism to translate their nucleotide sequences into amino acid sequences to make useful peptides. A droplet that happened to have the right mix of random peptides and RNAs for directing the synthesis of useful peptides would open up the protein world for evolutionary exploration. The problem is that it takes proteins to translate RNAs into proteins—a serious chicken-and-egg problem. No one knows how it was done, only that it was successful, since all cells now have this ability. We do not need to invoke miracles, but must carry out more experiments looking for small steps that could have been plausibly interconnected to bootstrap the whole process.

The system might have gotten going when peptides with an affinity to some but not all amino acids also had the ability to associate with small RNAs. If these double-duty peptides were in the same droplet with RNAs that lined up the small RNAs, so that the amino acids reacted with each other to form short chains with the same amino acid sequence as the double-duty peptides, an autocatalytic loop would be formed. The peptides would assist in making more of themselves. They would not have to be as highly specific for each amino acid as they are now, just good enough to tip the balance. The double-duty proteins that attach an amino acid to a transfer RNA (tRNA) in present-day cells are called aminoacyl-tRNA synthetases, and they are the secret to translation. In each organism there are at least twenty of these enzymes, one for each amino acid and its cognate tRNA. They can be assigned to one or the other of two families on the basis of their amino acid sequences. Comparison of these enzymes in a wide variety of living organisms makes it clear that the members of each family descended from a single sequence that was present long ago. The two families are unrelated and carry out their roles in slightly different ways. It seems that the molecular patriarchs of each gene family arose independently and have been fruitful and multiplied ever since.

There are many other steps in protein translation that might have been carried out initially by RNA molecules and only gradually replaced by proteins as random sequences happened to encode useful enzymes. They did not have to be very good enzymes at the beginning, just a bit better than the RNA they replaced. Natural selection would give a huge advantage to any droplet with an improved mix. A protocell with fifty or a hundred useful proteins could grow and replicate and fill the seas (Loomis 1988).

As protocells multiplied, they depleted the primordial soup and reached the first Malthusian limit. They could no longer rely on getting compounds from the environment and had to turn to another source of energy to be able to provide for themselves. They turned to the sun and became photosynthetic. The sun radiates 260 kilocalories of energy onto each square centimeter of the earth, thousands of times more energy than is released in electric discharges. All the droplets needed to do was find a way to harness some of the energy. Pigments that absorb light are present in the soup generated under a range of prebiotic conditions. Some of them bind metal ions in a sort of chemical cage that can absorb light so that the electrons on the metal atom get excited and can be passed to other molecules. By incorporating into the membranes some pigments carrying magnesium ions, protocells could pump protons out and generate an electrical potential that would bring protons back through an enzyme that made the universal chemical energy currency of all cells, ATP. With a few improvements and subtleties, this is how photosynthesis still works. A protocell freed of the need to scavenge for energy from the environment would have a selective advantage all over the world. The early bacteria fossilized in South African rocks show undeniable evidence that they were already using photosynthesis 3.2 billion years ago.

Somewhere along the line, RNA was replaced by DNA, which can be packed in a more regular fashion, as a double helix. RNA was still retained, but it was seldom replicated. New RNA molecules were transcribed off the DNA strands when needed and then broken down, so that the subunits could be reused. Some viruses might have been the first to use DNA as a way of avoiding the protection mechanisms that early cells had set up to avoid infection. Gradually, cells copied their own genes into DNA, so that they had both an RNA copy and a DNA copy, at which point there was no need to keep the RNA versions of the genes and they were all jettisoned.

Over evolutionary time, DNA has been a dynamic polymer, with deletions and duplications of regions going on all the time. When there are two copies of a gene, one can continue to carry out its original function and the other can change so that it encodes a slightly different protein. When new genes provide a selective advantage, they are kept in the population. In this way the number of useful genes increased until there were hundreds and then thousands of genes providing all sorts of useful functions. We would certainly call such a cell alive.

For the first billion years, the earth was bombarded by the debris left in its orbit around the sun. During this hellish period the surface was repeatedly broiled, parboiled, and generally left lifeless. The Hadean period ended about 3.7 billion years ago and the seas calmed. Within less than 200 million years, life arose and flourished in forms that resemble present-day anaerobic photosynthetic bacteria. Considering all the interwoven steps necessary for a cell to grow, replicate its hereditary information, and divide into equally adapted daughter cells, this is surprisingly fast. However, the building blocks were there for the taking; it was a matter of stringing them together. Not only were amino acids and nucleic acids generated on a variety of surfaces of the planet, some were made in space and delivered to the earth by interstellar dust, meteorites, and comets. Reactions that string them together in random order can be carried out in the lab under prebiotic conditions. The trick is to get copies of a specific order. Considering the number of independent droplets and the inherent errors in nucleic acid replication, the number of chances to come up with sequences that could catalyze accurate self-replication, as well as replication of other RNAs, makes the generation of an autocatalytic system almost inevitable. Expansion of such systems is limited only by their stability and the availability of necessary components. It is likely that any planet with the physical and chemical properties of the earth would generate life within less than 100 million years. We may not know the path or the details that generated the first living cells, but we do not need to invoke miracles.

THE OXYGEN REVOLUTION

The earliest cells had a life similar to that of modern-day anaerobic bacteria, since there was negligible free oxygen in the atmosphere (Knoll 2003). We know this from minerals, such as pyrite, siderite, and uraninite, which are destroyed by oxygen and yet are abundant in rocks laid down during the first 2 billion years. They remain there as buried testimony that the air would have quickly asphyxiated us. All the oxygen on the planet was initially bound in water, silicates, and other oxides. For 2 billion years, almost all cells lived by photosynthesis and the fermentation of organic compounds under anoxic, or oxygenless, conditions. Continuous selection for efficiency and accuracy of the enzymes necessary for this lifestyle perfected the genes and the interactions of the proteins they encoded. The anaerobic bacteria became some of the most highly evolved cells ever on the planet. However, most were killed when exposed to oxygen, which now makes up about 21 percent of the air we breathe.

Photosynthesis gives off oxygen and provides the energy to incorporate carbon dioxide into the reduced organic matter of cells. However, when a bacterium dies, its reduced matter is oxidized by other organisms, and all the oxygen originally generated photosynthetically is used up. In the long run there is no net change in the total amount of oxygen. Moreover, oxygen is used up in the weathering of rocks, so that the planet acts as an enormous sponge soaking up free oxygen from the atmosphere. For the first billion years of life on earth, oxygen accumulated only locally and was soon depleted as iron ores rusted and rocks weathered. The balance started to change about 2.2 billion years ago as huge numbers of bacteria trapped in seafloor mud were buried without being oxidized. Over millions of years, large regions of the ocean floor were subducted and pushed kilometers below the surface as tectonic plates collided. All of the trapped organic matter was then protected from oxidation for eons. Moreover, some bacteria generated copious amounts of methane, which we call "swamp gas." Methane rises high in the atmosphere, where it is broken down by ultraviolet light, generating hydrogen gas in the process. The hydrogen escapes into space,

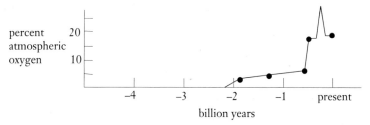

FIGURE 8.1 Accumulation of oxygen in the atmosphere. Although oxygen levels fluctuated significantly during the last billion years, only in the last 500 million years did oxygen reach levels we could live with.

leaving oxygen behind. This made an irreversible difference to the planet. It would never be anoxic again.

About 2.2 billion years ago, the level of oxygen in the atmosphere reached about 1 percent, and it steadily rose to reach its present-day level of 21 percent. The anaerobes that could not tolerate the presence of this strong oxidant retreated into the anoxic muds found at the bottoms of ponds, lakes, and seas. They are not gone, just out of sight. A few lucky bugs had systems in place that allowed them to survive oxygen and even use it to their advantage. Oxidative metabolism is much more efficient than fermentation, and the waste product, carbon dioxide, can be blown off. These aerobic bacteria took over the surface of the planet and filled many new environments.

About 1.5 billion years ago, a well-adapted aerobic bacterium that had been engulfed by an anaerobic archaebacterium set up shop in its host and provided the means for both of them to survive and thrive in the increasingly oxygenated world. Over time the aerobic bacterium passed most of its genes to the chromosomes of the host, which was living by that time as an aerobic eukaryote. The bacterium became the mitochondria of today (Margulis and Sagan 1995). Some of the descendants of the original eukaryotes lost their mitochondria, and a few became oxygen intolerant, but most went on to bigger and better things. They could pick genes from either of their parental strains and become specialized for one lifestyle or another. There are vase-shaped fossils in 750-million-year-old rocks of the Grand Canyon that are exactly the same size and shape as modern testate amoebae (Knoll 2003). These amoebae shared a common ancestor with *Dictyostelium* after the divergence of plants and animals, so we know when the major kingdoms arose. Plants picked up a photosynthetic bacterium along the way. This bacterium, a cyanobacterium, formed a stable alliance as a chloroplast, the photosynthetic organelle that provides energy for carbon assimilation in return for nutrients and protection. Plants spread over the surface of the land, changing the environment as they went. Amphibians, reptiles, birds, and mammals followed. The earth began to take on the appearance it has today.

Evidence that all eukaryotes are connected by common descent from a single ancestor is written in their genes. As mentioned in chapter 1, the gene encoding a specific metabolic enzyme is almost the same in humans, monkeys, chickens, coelacanth fish, insects, *Dictyostelium*, yeasts, and plants. When the amino acid sequences of this enzyme in different species are lined up, it is evident that almost half the positions have the same amino acid, whether the protein came from a mammal or a fungus. Neither chance nor convergent evolution could possibly account for this degree of similarity; only descent from a common ancestor explains it. Similar analyses carried out on thousands of other sets of related proteins have further reinforced this conclusion (Olsen and Loomis 2005). If we are all descendants of a common ancestor, where did all the diversity we see around us come from?

Most eukaryotic organisms are multicellular, and they build specialized organs that we can see. The exceptions are the early-diverging eukaryotes, the yeasts, and some algae that have retained their unicellular lifestyle. Multicellularity has evolved independently at least three times since the divergence of plants and animals. The mechanisms that hold cells together in a dividing egg are different in plants and animals, and *Dictyostelium* generates multicellular fruiting bodies in its own idiosyncratic way following the aggregation of hundreds of thousands of cells into a mound. The genes necessary for multicellularity were all there in the common ancestor of plants and animals; they were just used for other purposes.

Shortly before insects diverged from the line leading to vertebrates, a regulatory gene, the HOX gene, duplicated several times to generate a set of closely related proteins that control the expression of other genes. The HOX genes are considered to be master control genes, since they regulate when and where a diverse set of important genes function in the construction of segments and parts of limbs. A change in the function of a HOX gene can have far-reaching consequences, such as a reduction in the number of legs—from twenty in brine shrimp to the six we see in insects (Ronshaugen, McGinnis, and McGinnis 2002). A few mutations in one of these genes can generate not only new species but even new orders. The fossil record indicates that a species may remain morphologically unchanged for millions of years and then be replaced by a different species within ten thousand years, almost instantaneously on a geological time scale. Such punctuated evolution, unanticipated by classical Darwinian evolution, is the result of critical changes in control genes. Throughout the long period of morphological stability, there were gradual changes in the DNA sequence, but they made little difference to shape or size. Then one particular DNA change occurred that increased a species' fitness to altered ecological conditions, and the consequences showed up in the fossil record.

Other times we can be fooled by the apparent similarity of the embryonic stages in the construction of related but distinct species. When the head and tail are being established in embryos of a certain insect, the nuclei at the front are programmed differently from the nuclei at the posterior because a regulatory protein is present at the front but not the back. This beautiful explanation was thought to apply to all insects in which early embryogenesis seemed to unroll in the same manner. However, it turns out that the protein found at the front of a fly is different from the protein found at the front of a beetle. Surprisingly, the end result is about the same. The front still becomes the head.

The common heritage of vertebrates is easy to spot. Rocks laid down at the dawn of the Cambrian period 543 million years ago carry well-preserved fossils of animals with a stiff rod up their backs called the notochord. The nerve cord lies just above it and ends in a muscular tail. These chordates gave rise to the vertebrates, with their characteristic backbone. The gradual replacement of the notochord by the spinal column during evolution is recapitulated during the stages of early vertebrate embryogenesis: the notochord forms the disks between the vertebrae, and the dorsal nerve cord becomes the brain and spinal cord. At about this stage, all vertebrates develop pharyngeal pouches just behind their heads, which make the embryos of fish, reptiles, birds, and mammals look surprisingly alike. These pouches become parts of the gills in fish, but they are lost during later stages of animal embryogenesis.

By chance the HOX gene cluster was replicated twice in a fish genome, and the extra copies helped pattern fins into limbs. The arrangement of bones in all limbs, going from the shoulder to the fingers, and from the hip to the toes—one bone, two bones, many bones—exists in certain fish and all four-legged animals (Daeschler et al. 2006). Fossil evidence tells us that this is an old

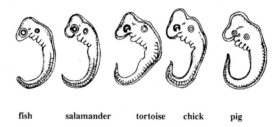

| fish | salamander | tortoise | chick | pig |

FIGURE 8.2 Embryonic forms of amphibians, reptiles, birds, and mammals look surprisingly alike when the pharyngeal pouches form just under the head. All mammals, including humans, have a pronounced tail at this stage. Redrawn from Ernst Haeckel, *Generelle Morphologie der Organismen* (Berlin: Georg Reiner, 1866).

pattern that has never changed. During the Devonian and Carboniferous periods, between 400 and 300 million years ago, both fish and amphibians, some as big as four meters in length, lumbered around on legs with the "one, two, many" pattern of bones. They had to lay their eggs in water but were otherwise kings of the land.

At the end of the Carboniferous period, as the land dried out, a species of amphibian produced a rare mutant that put a waterproof covering on its eggs, which then did not dry out when they were left on land by mistake. The hatchlings evolved into the reptiles that took over the earth for more than 200 million years. Of the amphibians, only the small ones that could hide near streams survived. During the Triassic, Jurassic, and Cretaceous periods, 248 to 65 million years ago, the reptiles diverged into a myriad of species, both large and small, some with feathers and some with hair. Over time, random mutations in one of the small hairy ones changed the way that the shoulder and hip bones developed, so that the legs ended up underneath the body and the stance became more upright. One of these evolved the ability to feed its young with mothers' milk and gave rise to mammals. Initially, they laid eggs, as the duckbilled platypus still does, but later variants allowed the embryos to develop internally. By a hundred million years ago, placental mammals were in place and ready to go once the dinosaurs went extinct. During the last 65 million years they have generated aardvarks, armadillos, bats, beavers, porcupines, and primates.

THE EVOLUTION OF HUMANKIND

Although we should not think of ourselves as more evolved than any other living species, we are naturally interested in our own family tree. What did our forebears look like, where did they come from, and what did they do? The fossil record and genomic analyses of DNA in populations all over the world have filled in the answer in the last twenty-five years. Early primates such as lemurs and tarsiers have been around for over 60 million years. They differ from their shrewlike insect-eating ancestors in having nails rather than claws at the ends of their fingers and toes. About 35 million years ago, some primates evolved into monkeys about the size of fat squirrels, which ate fruits and nuts. Over the last 30 million years, as the Atlantic Ocean opened up, Old World monkeys evolved separately from New World monkeys. In the Old World (Africa), the line led to lesser apes, such as gibbons, about 23 million years ago. Orangs split off the line of apes about 15 million years ago, followed by the gorillas 9 million years ago. Hominids and chimps separated only about 5 million years ago. About 4 million years ago our distant ancestor *Australopithecus afarensis* evolved from *Ardipithecus ramidus*, which had apelike teeth and lived in the Afar region of what is now Ethiopia (White et al. 2006).

Lucy is the most famous representative of *A. afarensis*. Her fossilized skeleton, discovered in 1974, clearly shows that she walked upright. Her leg bones were adapted to holding her whole weight, and her pelvis had shifted relative to those in great apes that move on all fours. She was a young adult when she died 3.18 million years ago. She stood three and a half feet tall, weighed about sixty pounds, and had a protruding, apelike face with a square jaw. Fossils of over three hundred other individuals of *A. afarensis* have been discovered, but Lucy's skeleton is the most complete. A footprint of a member of this species was found preserved in wet volcanic ash that fell 3.6 million years ago. It shows that *A. afarensis* walked the way modern humans do: heel first, followed by a toe push-off. *A. afarensis* is the only ape known to have walked upright at this time, making it almost certain that all subsequent bipedal hominids, including humankind, are descended from this species.

The fossil record for *A. afarensis* runs out about 3 million years ago, but a new species turns up in South Africa and Tanzania. *Australopithecus africanus* fossils have been found in rocks dated to 2 and 3 million years ago. The brains of these hominids were larger than those of *A. afarensis* and, at 485 cubic centimeters, were well above the average size of brains in modern apes. However, their arms hung down to their knees in classic ape-man fashion. At least three other hominid species appeared toward the end of the period in which *A. africanus* lived. The faces of two of them are dramatically more robust, but these species are probably not in a direct line to *Homo sapiens*. Their skulls, cheekbones, jaws, and teeth were modified to allow the attachment of strong muscles adapted to eating large amounts of tough, fibrous foods. The three other hominid species appear to have gone extinct when the climate changed about a million years ago.

Fossils of *Homo habilis*, the other species that replaced *A. afarensis*, were first uncovered in the Olduvai Gorge of northern Tanzania by Mary and Louis Leakey in 1964. They immediately realized they had stumbled upon a species more closely related to *H. sapiens* than to other hominids. *H. habilis* had a larger brain than *Australopithecus* and had mastered the art of fashioning stone tools. Similar but distinct fossil remains from the Olduvai have been assigned another name, *Homo rudolfensis*, and it is not clear which one might have been ancestral to humans. Perhaps both, if they were actually variants of the same species and could still mate with each other and produce viable offspring. Both were small and had long, apelike arms. Their fossils have been found associated with stone tools at multiple sites in East Africa. About 1.5 million years ago, one or the other, or both, gave rise to *Homo erectus*, who had an even larger brain—1,000 cubic centimeters—and teeth that resemble our own; *H. erectus* stood five to six feet tall. Some members of this species walked out of Africa and spread all over Asia. Fossils of forty individuals were found in the 1920s in caves at Zhoukoudian, fifty kilometers southwest of Beijing, China, and were originally identified with the name Peking Man. The species appears to have lived communally in big caves in the region for over 300,000 years. Layers of ash and charred animal bones on the floors of the caves attest to their mastery of fire. They brought wood into their caves for cooking dinner, warding off the cold, and keeping predators away. Toward the end of their stay, about 400,000 years ago, they had learned how to fashion clothes with bone needles and adorn themselves with necklaces of animal teeth.

An equal number of *Homo* fossils found in Java were identified with the name Java Man. Only later, following further finds in Africa and Europe, did paleoanthropologists realize they were all the same species, *Homo erectus*. The species must have been smart and adaptable, since they survived for more than a million years. The last descendants appear to have lived until 12,000 years ago in isolation on the Indonesian island of Flores (Morwood et al.

2005). However, it was back in Africa that they gave rise to *Homo neanderthalensis* and *H. sapiens*. Fossils of *H. erectus* found in Africa had traits more similar to those of modern humans than those seen in members of the species that had adapted to life in Asia and Europe. By 1.6 million years ago, the African hominids had learned how to make large cutting tools such as hand axes and cleavers. Some paleoanthropologists refer to them as *Homo ergaster*. Cultural development continued in Africa, and now and then a group went wandering into Europe but did not leave many remains.

The Neanderthals were the first to leave Africa and successfully populate the Middle East and Europe. Their remains have been found in the vicinity of Qafzeh Cave, Israel, as well as in Spain and Uzbekistan, at sites dating from 200,000 to 25,000 years ago. They looked almost the same as modern humans but had sloping foreheads with beetle brows, and their bodies were broad and stocky. As mentioned earlier, the size of their brains was about the same as that of modern humans, and there is no reason to think that they were slow witted. They buried their dead with ceremony and traded for needed items. Although it is possible that the arrival of *H. sapiens* led to their demise about 25,000 years ago, they may have been on their way out for other reasons. In any case, *H. sapiens* came out of Africa, and nothing seems to have stopped their expansion ever since.

The oldest remains of modern humans have been found in Africa and dated to 130,000 years. The ancestry of everyone alive today can be extrapolated to a single woman who lived at about that time, the Mitochondrial Eve (Cann, Stoneking, and Wilson 1987). She had telltale identifiers on her mitochondrial DNA, which was passed down to daughters, granddaughters, and so on in a matrilineal line to today. Fathers provided chromosomes but not mitochondrial DNA. As this woman's descendants spread out over Africa during the next 50,000 years, they inherited her unique mitochondrial markers and, now and then, added further harmless mutations that can now be seen in certain populations in Africa. There were other women alive at the time of Mitochondrial Eve, but they either had no daughters or their mitochondrial lines ran out when all their descendants were male.

By using the sequence of the male Y chromosome, the paternal line of all living humans has been traced to a single man who also lived in Africa, about 100,000 years ago. All the variously shaped and shaded people of the earth are descendants of these African hunter-gatherers. The ancestral markers turn up most often in the San people, who now live in South Africa; the Biaka Pygmies of central Africa; and certain East African tribes, but are present with modifications in every population tested, including Arabs, Australian Aborigines, Caucasians, Chinese, and Native Americans from Alaska to Tierra del Fuego.

When the mitochondrial chromosomes are analyzed in a large number of diverse peoples living outside of Africa today, the markers indicate that a small band of only about a thousand individuals from the Horn of Africa migrated to the Middle East and along the Persian Gulf about 60,000 years ago. They may have encountered dwindling populations of Neanderthals who had left Africa 100,000 years earlier but, if so, soon replaced them. By following the rare mutations that occurred now and then on the way, we can clearly see the paths taken by these wanderers. A few isolated groups, such as those on the Andaman Islands off the coast of Myanmar, have retained the early markers almost undiluted by later variations. Some markers are found only among the Australian Aborigines in their isolation on the southern continent. Amazingly, human remains that date to 50,000 years ago have been found near Lake Mungo in southern Australia. Within 10,000 years, descendants of that first small band of hunter-gatherers spread along the coast of India, down the Malay Peninsula, and across the islands of Indonesia to Australia. It seems they never turned back.

At the same time, other branches of the family spread into Europe, where their 36,000-year-old artwork can be seen on the walls of caves in southern France. Yet another branch skirted the Himalayas and moved through central Asia to reach the Yalu River in Siberia 30,000 years ago. They stayed there for about 10,000 years, until a land bridge opened between Siberia and Alaska as a result of a drop in sea level. Humans who had lived in the Altai region of Siberia walked across and somehow managed to move down the frozen coast to the verdant plains of North America. They were met by immense herds of bison and mammoths that provided a rich source of meat. They must have spread out rapidly, since they left unmistakable remains at the Meadowcroft Shelter in Pennsylvania that are 16,000 years old. Others moved south through central America, the Amazon Basin, and the Andes, reaching Monte Verde in southern Chile 14,000 years ago. Within 40,000 years *H. sapiens* had filled almost every corner of the world. The last areas to be settled were the islands of Polynesia. Since then, the species has multiplied to our present world population of 6.6 billion people.

The twenty-thousand-kilometer trips taken by our ancestors did not have to be a series of quick marches. If, at some point during each generation, some of the settlers moved a day's walk down the beach or rowed forty kilometers along the coast, the species would reach the ends of the earth in five hundred generations, or about 10,000 years. The real question is why these bands moved out of Africa and kept going. Their cousins were content to stay in Africa, where they knew the land and their neighbors. Perhaps the wanderers came from a tribe in which there was a high prevalence of a rare abnormality—adventurous curiosity. Although we do not know the neurological basis for curiosity, most would agree that we are a curious sort. Perhaps a rare mutation that resulted in a slight rewiring of the brain was fixed in the population of a village back in Africa 60,000 years ago and produced a feeling of greater reward when curiosity was satisfied. Curiosity is a dangerous but wonderful thing. Without it we would not have science or questions about our origins.

Along the way, the wanderers diversified into hundreds of groups that can each be distinguished by their genetic makeup, size, language, skin tone, and hair. There is more diversity in Africa than the rest of the world combined. The earliest branches stayed in Africa and adapted to high arid plains, lush tropical jungles, and burning deserts. Some tribes in Egypt, Sudan, and Ethiopia carry more of the ancestral markers found in non-Africans, but they also share genetic markers with neighboring Africans that are not found in populations of Europeans or Asians. Those in the northeast of Africa are more closely related to the small group that migrated out of Africa 60,000 years ago than to other Africans, but there has been mingling of the populations over the years as famine or wanderlust drove people to new places.

Little is known of the habits and culture of the early colonists. They appear to have lived in groups of twenty to forty individuals and cared for the elderly and infirm. They adorned their dead with precious objects and buried them ceremoniously. Most important, they left paintings on the walls deep inside caves that give us tantalizing glimpses into their lives and worldviews. The animals they hunted take center stage; the hunters are only rarely depicted. The similarity of styles in caves along the Dordogne River in France and the Altamira cave on the north coast of Spain suggests that a common culture flowed over southern Europe 25,000 years ago. When I visited the cave at Peche-Merle near

Bordeaux, I was amazed by how similar were the forms preserved on the walls and those I had seen 20 years earlier at Altamira. The painters almost seemed to have gone to the same art school. The use of the rock walls, the size of the animals, the colors, and the shading of the outlines of bulls, horses, and deer were almost identical.

Cultural exchange does not require an invasion. New and interesting ideas can be transmitted by a few visitors from far away and then spread within the indigenous people. There is some evidence that this may have occurred when voyagers from Oceania became lost and their ocean-sailing canoe landed in northern Chile centuries before Incas and Europeans came down from Peru. The ceramics and pottery found near the village of Diaguita just south of the Atacama Desert show motifs surprisingly similar to those prevalent in Oceania. It could have been a chance convergence of styles, but it is more likely the result of the locals' copying new designs they found attractive. The seafarers were too few to leave any genetic trace, but their art was passed down.

In the last few hundred years, we have learned to travel the world and the seven seas. Adventurers and tourists have returned with trophies, knickknacks, and stories of cultural traditions that challenge our own ideas of what is normal. Sometimes contact between civilizations leads to violent reactions, but more often to healthy curiosity. Global communication makes the whole world a village where we know when someone succeeds and someone fails almost as soon as it happens. Recognizing our common history and heritage will help us all get on.

Continuing to live together on this planet is going to take a major effort. We are so successful as a species that we have multiplied beyond bounds and pushed up against the absolute carrying capacity of the earth. If we want life as we know it to continue, we are going to have to limit our use of dwindling resources and stop polluting the world around us. Life is more than breeding to the maximum. It requires thinking ahead and tailoring our actions and urges to the commons we all share. Putting life back in balance is the subject of the final chapter, perhaps the greatest challenge we have ever faced.

NINE

SUSTAINABLE LIFE

AS A MICROBIOLOGIST I HAVE GROWN COUNTLESS FLASKS OF BAC-
teria in the laboratory. When the initial number of cells is low, the
bacteria double every hour and the population increases. As they
fill up the flask, nutrients become limited and wastes build up so
that growth slows and then stops. The next day most of the bacteria are dead. It is best not to anthropomorphize too much about
bacteria when you are an experimental scientist, but it is impossible not to think that the fate of any organisms that follow such a
course might be the same, whether they are wasps, lemmings, or
humans. When the wasp population gets above the carrying
capacity of the environment, many of them die but a few fly away
to start another nest. Lemmings can go on mass migrations,
though some die along the way. But where can humans go?

Studies of nonhuman populations usually find that, once the
carrying capacity of an ecosystem is exceeded, a severe crash of
the population occurs in association with rapid environmental

FIGURE 9.0 *Picking valuables, Cambodia.* Photo by Maciej Dakowicz.

degradation. Scientists watched a dramatic example of such a boom-and-bust cycle on a small island in the Bering Sea off Alaska. In 1944 a herd of twenty-nine reindeer was brought to Saint Matthew Island, an uninhabited spit of land thirty-two miles long and four miles wide. It was reindeer heaven, with lichen mats four inches thick covering much of the island, and no wolves or other predators, just birds and some foxes. By 1957 there were over fourteen hundred reindeer, most of them fat and in excellent shape. By 1963 there were six thousand reindeer crowded forty-seven to the square mile. The lichen mats were badly trampled and overgrazed. Perhaps the most ominous sign of an imminent population crash was the scarcity of yearlings. Three years later only one sick male and forty-one females remained alive. The herd had overshot the carrying capacity and degraded the landscape so badly that it could not recover before the reindeer had completely died out. We humans have more foresight than reindeer, and we should be able to control our population before we reach the tipping point.

How Many People Can the Earth Support? is the title of a seminal book by Joel Cohen published in 1994. It does not answer the question, but instead gives a set of answers that depend on how the question is taken (Cohen 1994). There is a big difference in the maximum number of healthy, well-fed, happy people the earth can support and the maximum number of sick, famished, and desperate people. Estimates of the human carrying capacity of the planet have been based on the total amount of energy available, the amount of food, the net primary productivity of photosynthetic plants, and something called the ecological footprint. Estimates made over the last forty years have ranged from 1 billion to 14 billion, and the median estimate is 3 to 4 billion people. The basic resources of the planet such as land, water, and energy are limited. Moreover, humans are a restless, energetic

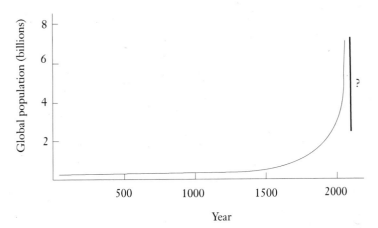

FIGURE 9.1 Boom and bust of a population. The human population on the planet was constant until the sixteenth century, when it started to increase exponentially.

sort that do not always live in close harmony, especially when resources are limited. Population fluctuations due to war, famine, and pestilence have to be considered when estimating the carrying capacity of the earth. How stable a population are we talking about? How degraded an environment are we willing to accept? Are we talking about the earth as we now know it, with abundant reserves of fossil fuels, or are we extrapolating to the future, when they are exhausted? Most people who care about the lives of their grandchildren and subsequent generations are interested in sustainable life.

The human population worldwide was less than half a billion before A.D. 1600. It doubled in the next two hundred years and doubled again in the next century (Cohen 1994). A hundred years later, in 2000, it stood at about 6 billion. In the next six years it increased to 6.6 billion. Depending on which of the best estimates of the carrying capacity you accept, we are either at the maximum for the planet or considerably above it. Unfortunately, it is estimated that the population will continue to increase to at least 8 billion even if all women restrict their families to an average of 2.1 children, the number that replaces them and their partners. Over 1.8 billion people on the planet are under the age of fifteen, and most of them have not yet started to add their children to the count. The danger of a population crash is very real.

The problem is compounded by the unequal distribution of the wealth among nations. People in the developed world make up 20 percent of the total population yet consume 86 percent of the world's assets. The rest of humanity lives on the remaining 14 percent. Moreover, most of the increase in population is happening in the underdeveloped nations: 95 percent of all the people being added to the world's population live in the poorest countries. The population density is more than twice as high in the underdeveloped countries as in more developed, richer countries. Given the chance, many people in the poor, crowded slums of the world would limit their families. They know the consequences of adding more children and would like a better life for their children. When family planning and birth control are provided, rates of population increase drop rapidly. If there is a technological fix to the population problem, perhaps it lies in bringing better family planning to the poor of this world.

BETTER LIVING THROUGH BIOLOGY

When humans lived as roving bands of hunter-gatherers, there was no overpopulation or ecological degradation. A few hundred thousand humans could live in harmony with nature and leave only footprints. Starting about ten thousand years ago, the agricultural revolution changed all that by providing a large and dependable supply of food that allowed people to live settled lives near their fields. The first farmers probably did little more than collect seed and spread it near their huts. Perhaps they weeded their plots and watered the plants at times to increase the harvest. The invention of the plow and wheel allowed large-scale farming and further increased agricultural production.

Other agricultural innovations included the domestication of sheep and other farm animals. There is evidence that more than eight thousand years ago mouflon in the mountains of Persia were captured and their lambs raised in enclosures. Selective breeding for docility and fecundity produced large herds that could graze on the hills and be brought back to the village to provide wool, milk, and meat. Shepherds led a primitive life tending their sheep in the woods and fields, but a life that was far more civilized than that of hunter-gatherers. Wealth in some parts of the world is still measured by the number of domestic animals held by the family or tribe. A wealthy tribe can better care for children and expand. Food surpluses freed some farmers to do other work, such as arts and crafts, which they traded with neighboring tribes. Surpluses allowed societies to grow until they reached the effective carrying capacity that the technology made available. When the land was depleted, they could move on, although they might have to displace weaker tribes to gain access to good land.

The agricultural revolution proceeded to make advancements for thousands of years as new crops were brought under control and sophisticated irrigation systems developed to make the desert bloom and fill the rice paddies. Most farmers are conservative and stick to the traditional ways, but there are always a few who will try out new ways to sow and reap, to breed and keep, to fertilize and irrigate. When a farmer is successful, his neighbors know about it immediately and copy the style. Harnessing a horse or a cow to the plow was once a brilliant new idea that allowed the farmer to cultivate much more land. It is now taken for granted that a horse, a bullock, or a tractor will pull the plow, and fewer farmers can produce enough food for everyone.

Intensive agriculture allowed some populations to expand dramatically. However, historical records of periodic famines in different parts of the world attest that the population has always been close to the effective carrying capacity. In 1943 the worst recorded food disaster occurred in eastern India. An estimated 4 million people died of hunger that year during the Bengal famine. India has been haunted by the memory of this terrible event and has made huge efforts to avoid letting it happen ever again.

Even though the world's carrying capacity gradually increased as a result of improved agricultural methods, the population was never far behind. The second half of the twentieth century saw a partial respite from massive famines as new breeds of crops were introduced. For the first time, crop breeding became a science and not just a trial-and-error, empirical process. Gregor Mendel's laws concerning the inheritance of traits, discovered in relation to sweet peas, were found to hold in other plants too. In 1926 Thomas Hunt Morgan published a book titled *The Theory of the Gene*, in which he laid out the experimental data supporting Mendelian inheritance in all sexually reproducing organisms. But the "gene" was still only a theory in Morgan's head. Within the next twenty-five years, work with viruses, molds, corn, fruit flies, and mice clearly defined genes and chromosomes and showed how they controlled traits. In 1953 the chemical basis for hereditary information was shown to be encoded in the sequence of bases in the double helix of DNA. For the first time, plant breeders could think sensibly about dominance and recessiveness of genes, linkage, mutation, and balanced stabilization. Large-scale breeding efforts were set up to improve varieties of wheat, rice, corn, and other crops.

The Green Revolution developed wheat cultivars with shorter, wind-resistant stalks, as well as fast-growing varieties adapted to cold climes or the tropics. In the far northern plains of Canada and Siberia, seeds that germinate a few days earlier in the spring have a much better chance of being harvested before the first freeze in the fall. In warmer climates new varieties were used to give two harvests a year in heavily fertilized fields. Over half the corn now being grown uses the modern varieties, and wheat production worldwide has increased fiftyfold since 1950.

Massive efforts were also applied to breeding better rice. Seeds were collected from all over the world and used to map quantitative traits. They were mixed and matched to produce high-yield strains for all seasons, all soils, all needs. By 1980 the growing populations in India, Indonesia, and Asia were able to feed themselves. Between 1967 and 1977, India went from being a chronically food-deficient country to being one of the world's leading agricultural nations. The land under cultivation increased, double crops could be expected most years, and the crops of rice, wheat, millet, and corn gave 30 percent higher yields. Irrigation systems were built to provide water during the dry season, and fertilizers and pesticides were added to the fields. For the first time in a long time, Indians did not face famine.

For the last thirty years, famines have been a problem not of production but of distribution. Grain can be shipped or airlifted to isolated regions suffering from drought or blight. Only political upheaval or incompetence has led to serious hunger. During this period the worldwide food supply has been sufficient to adequately feed everyone on earth. But at a price.

Growing high-yield crops requires a heavy input of fertilizers, and herbicides and other pesticides, which over the years can poison the surroundings. Prolonged irrigation in hot, dry regions leads to the gradual buildup of salt as the water evaporates. Excess water can percolate into groundwater and raise the water table so that it seeps into the plant root zone carrying dissolved salts previously stored in the soil and rock. Evaporation during the dry season can leave salt on the surface, reducing the ability of plants to extract water from the soil. Some salts may be toxic to plants, further reducing their vigor and growth. Crops will be stunted and pastures will show patchy growth. More than a million acres of croplands in the Central Valley of California have become highly saline in the last ten years and are no longer suitable for crop production. Plans to stabilize the valley for sustainable agriculture include discharge of drain water into the ocean or concentration in local basins. It will take massive engineering to bring this land back into balance with nature. Together with the high price of fertilizer and mechanization, these shifts in the agricultural economy have favored large industrial farms over small farmers and have led to greater inequality between rich and poor farmers. It has been a very high price to pay.

In the last ten years, agricultural production has benefited from advances in understanding the DNA basis of genes. Techniques have been developed for taking a gene out, modifying it in the lab, and putting it back into chromosomes to make genetically modified (GM) plants carrying desired traits. As mentioned in chapter 3, a bacterial gene for Bt toxins is being used to kill insects and worms. Plants that make their own Bt proteins can be grown with far less use of pesticides and resist attacks by corn borers, boll weevils, nematodes, and other pests. This protects the land from pesticide buildup without reducing the yield. GM crops have also been engineered to resist specific plant viruses, thereby improving production. Others have been made salt tolerant and able to flourish in the desert. However, the dangers of sowing a large number of fields with genetically identical seeds year after year are well known. Monocultures are always in danger of catastrophic failure if the weather is abnormally wet or dry. The specific cultivar may be resistant to worms and fungi but may perish for lack of water. Moreover, parasites and pests adapt to the particular cultivar and can wipe it out in a few years. Many of the problems of GM monocultures can be avoided by using different outcrossed strains each year and making variants available to local farmers. There are other environmental worries about GM crops, but they are outweighed by the obvious benefits.

GM food crops now include corn, cotton, soybeans, canola, squash, and papaya grown in the United States, Argentina, Canada, Brazil, China, and South Africa. The number of fields planted with GM seeds increases annually as the benefits to the farmers and the national economy become increasingly apparent. Half the corn and almost all the cotton and soybeans in the United States are now grown from GM seeds.

Back in 1995 I was an expert witness for a patent on genetically modified tomatoes that can ripen on the stem and not rot on the way to market. These "Flavr Savr" tomatoes had been engineered by the agribusiness Calgene to express an antisense RNA that blocks synthesis of an enzyme that turns them to mush. It seemed like a good idea to me. I testified that the technique was novel in plants and provided a useful product. The patent was upheld. Similar techniques have now been used to keep strawberries, pineapples, peppers, and bananas fresh for longer.

In some parts of the world, diets are deficient in vitamin A. Golden rice has been engineered to contain lots of vitamin A by moving two genes from daffodils and a third gene from a bacterium into the rice chromosomes. Vitamin A plays an important role in vision and many aspects of cell growth and differentiation. It also helps regulate the immune system and acts as an antioxidant. Golden rice has improved the health of millions and prevented blindness in hundreds of thousands of children in developing countries every year since its introduction in 1999. In 2005 the strain was further modified to accumulate twenty times more vitamin A and is now growing in India and the Philippines (Paine et al. 2005). This is better living through molecular genetics.

Nevertheless, people in many countries do not trust these new GM crops and fear that they might be unhealthy in the long term. Although there is no evidence that plants modified by molecular genetic techniques are any different from crops generated by natural crosses, GM crops are strictly controlled in much of the world. Yet there is always the danger that they will inadvertently spread out of designated fields and generate volunteers that can take over adjacent fields. Partly to prevent such spread, GM crops have been developed that generate only sterile seeds. Various "terminator" techniques have been used to make the seeds of the first harvest unable to generate new plants. Some carry a gene that produces a toxin when the seeds from the first crop are planted. The toxin kills the second crop. Others carry a mutation that makes them sterile unless treated with a chemical available at the biotech company where they are grown. Such terminator seeds cannot spread. They also protect the biotech companies from unauthorized use of their products that could jeopardize their chances of recovering the cost of producing improved seeds.

And therein lies the rub. New seeds have to be purchased each planting season. It has been argued that terminator seeds will turn small farmers into "bioserfs," destroying their age-old custom of saving seeds for next year and making them completely dependent on the biotech companies. In 1999 Anuradha Mittal and Peter Rosset of Food First wrote, "Patenting genes the same way you patent software robs Third World farmers. While they and their ancestors developed almost all important food crops, transnational corporations can now blithely patent those crops and make mega profits without in any way compensating traditional farm communities for the original research. Genetic resources taken freely from southern countries will be returned to them later as pricey patented commodities. 'Terminator' technology is a way of locking this 'bio-piracy' into the very genes themselves."

As the result of mounting protests from developing nations, Monsanto, one of the largest producers of GM seeds, vowed to keep its terminator seeds off the market. As recently as 2006, five hundred thousand people in India signed a petition calling on Prime Minister Manmohan Singh to uphold the country's ban on terminator seeds, and the European Parliament reaffirmed its support for a moratorium on terminator technology. It has become a politically charged matter that further clouds the use of genetically modified crops. In any case, the biotech companies continue to produce and sell GM seeds with many beneficial traits that have helped to feed the hungry and reduce the buildup of pesticides.

Fish are another major source of food. Fishing has been a way of life for thousands of years. The bounty of the seas once seemed inexhaustible, but drift nets and one-hundred-kilometer longlines have effectively scoured the seas in the last fifty years. Huge floating factories pull in what fish are left in the oceans and process them as they go. With 90 percent of the big fish gone, the days of fishermen as hunter-gatherers bringing their catch from the seas to market are numbered. The catch of tuna, swordfish, marlin, cod, halibut, and flounder is a tenth of what it was in 1950, and those that are caught are less than half the size they used to be. Smaller, less desirable, fish are now under siege, and their numbers will soon decline. Yet fish provide the only source of animal protein for a billion people, and the rest of the world has an appetite that consumes 90 million tons of fish a year. Shortages are not apparent in the homes or restaurants of the affluent world, but in the small fishing villages of the developing world, people are going hungry.

Just as we no longer expect to be able to put meat on the table by going out to shoot it, we cannot continue to expect to put fish on the table that have been caught in the wild. Farm-raised fish presently make up to 40 percent of the fish that people eat. All around the world, pens are crowded with fish that are fed and protected from predators and poachers. Farming techniques for most fish are still somewhat primitive and result in local pollution and occasional mass escapes that can affect the wild population. Carnivorous predatory fish such as salmon have to be fed two to five kilograms of ground fish fry for every kilogram of body weight they yield, and there is a lot of waste. Tilapia, on the other hand, eat algae and plants. This warm-water fish has been farmed for thousands of years in the Middle East, where it is known as Nile perch or Saint Peter's fish. The fish are hardy, and their meat can be tasty. They grow rapidly in fresh or brackish water, and the wastewater can be used to fertilize nearby vegetable fields or paddies. They may provide a way to feed the masses.

Caring for crops, fish, and cattle to support the present population depends on cheap energy, mostly in the form of fossil fuels. It is used to pump irrigation water and mix the water in fish pens. And it is used to make and haul food and fertilizer and to bring products to market. But the easily recovered fossil fuels are running low, and gas will be prohibitively expensive in a generation or so. We need a sustainable source of energy to maintain the standard of living—the way of life—for both the developed and the developing countries.

One solution to the energy crisis may be to improve ethanol production from sugarcane and adapt automobiles and small engines to run on this fuel. When petroleum is history, we may get around on ethanol. Sugarcane stalks stand four meters tall in the bright sun of hot, humid fields. Their sucrose content is one of the highest of all known plants and might be increased further by applying molecular genetic techniques. Sugarcane molasses is fermented to ethanol, which is distilled and sent to the pumps. The waste is burned to generate steam for the turbines that run the ethanol production plant and provide all its energy needs. In Brazil, the leading producer of ethanol, 90 percent of the new cars run on a combination of ethanol and gasoline. Genetically modified sugarcane plants resist sugarcane mosaic virus and produce significantly more sucrose per acre. Other, more robust GM strains are under development. Sugarcane has also been selected to express a bacterial gene that converts sucrose to another commercially valuable sugar, isomaltulose, while still in the fields. Molecular genetics has only just begun to improve the overall yield of ethanol from sugarcane to prepare for the day when the gas runs out.

Biodiesel holds out even more promise. It can be made from wood chips, weeds, straw, garbage, or sewage sludge and yields twice the energy needed to produce it. If the raw material is produced on agriculturally marginal land, there would be little or no effect on food supply. Prairie grasses and roadside weeds might be harvested and locally converted to transportation fuel. At present, the high cost of rapidly breaking down the cellulose has made production of biofuels unprofitable, but progress is being made in tailoring enzymes for industrial-scale processing for biodiesel. Widespread use of cellulose-based biodiesel would significantly reduce dependence on fossil fuels.

GLOBAL POLLUTION

Burning petroleum, natural gas, and coal, along with deforestation, adds 7.7 billion tons of carbon dioxide to the atmosphere each year. This is only about 3 percent of the amount that naturally cycles through the atmosphere each year as the products of photosynthesis decay or are eaten, but it is a net addition to the planet that will not be removed until the global ecosystem changes radically. Carbon dioxide levels are now at 0.04 percent, higher than ever in the last 650,000 years, and the level is increasing each year. While it is not clear how much of this increase comes from human activity, most of the increase occurred in the last 200 years as the population increased sixfold, large areas were deforested to provide farmlands, and fossil fuels were burned in abundance. Before the industrial revolution, carbon dioxide levels were at 0.03 percent. The increase would not be a cause for concern if carbon dioxide were not a greenhouse gas. Together with methane and nitrous oxide, carbon dioxide absorbs much of the heat that reflects off the earth, working like the windows in a greenhouse to trap the heat.

And the planet is heating up. The annual average surface temperature worldwide in the last few years has been the highest since instrument recording started 200 years ago. Glaciers are melting at unprecedented rates in the Alps, in the Rockies, in Greenland, and on Mount Kilimanjaro. There are fears that large portions of the Greenland and Antarctic ice sheets will flow into the sea and raise the sea level by several meters. Sea levels will also rise as the water warms up and expands. After being fairly stable for two thousand years, mean sea level increased twenty centimeters in the last century and is predicted to rise another meter or more in the next century. This is not a good time to buy a beach house.

Some of the largest cities will be flooded and whole nations may be submerged. The highest point on the Maldives in the Indian Ocean stands only three meters above present sea level, and the 350,000 people that live on this chain of islands may have to move soon. It is unclear if we can do anything about the rise of the sea.

The climate on earth has gone through wild swings since long before humans made any difference. Every few hundred million years, glaciers have advanced from the poles to cover huge portions of the continents. The whole planet may have frozen over 600 million years ago, becoming Snowball Earth (Olcott et al. 2005; Kopp et al. 2005). Some of these ice ages lasted for millions of years before the glaciers retreated. On the other hand, temperatures soared about 55 million years ago as the result of a massive release of the greenhouse gases carbon dioxide and methane from the seafloor. During this period crocodiles lived on Ellesmere Island, the most northerly point of the North American continent. The surge in greenhouse gases raised the surface temperature ten degrees centigrade within a few thousand years, roasting those in the tropics. For the last million years, the earth has been in a cold period, with glaciers periodically advancing and retreating. The most recent ice age reached its height 20,000 years ago and ended about 8,000 years ago. We are now living in a warm interglacial period that is expected to end in 30,000 years when the glaciers advance once again. That is such a long way in the future that we need not worry about it now.

Meanwhile the climate is warming, and there are certain to be major changes in the weather in our lifetimes. The monsoon rains that India depends on for intensive agriculture may fail or follow a different track. Growing seasons may lengthen in the Far North even as the equatorial deserts expand. There are so many connections between the climate and other things—butterflies, jet streams, water flows, and so on—with some reinforcing changes and others counteracting them, that it is impossible to predict with any confidence what will happen. Still, it seems sensible to reduce the buildup of greenhouse gases, and efforts in that direction are under way. However, it may be too late, since changes on a planetary scale have such a long lag time that we are now seeing the consequences of events 50 years ago and will not see the consequences of our present actions for another 50 years.

Efforts to wean society from fossil fuels have included the use of solar, wind, water, and nuclear power. These do not put carbon dioxide into the atmosphere as they generate electricity, but there is a start-up cost in building the plants that is seldom brought into the equation. It takes a lot of energy to fabricate the huge modern windmills and even more to build dams for hydroelectric plants. And the energy comes from burning fossil fuels. Nuclear plants are not cheap to build either, and they have the added disadvantage of generating radioactive waste that has to be stored for thousands of years before it is safe to come near. We don't know what to do with it but bury it in a mountain and hope it stays there. If we rely on nuclear energy globally for most of our needs, we are going to leave a lot of radioactive waste for future generations to worry about. There does not seem to be an easy way out.

At the end of 2005, the United Nations General Assembly proclaimed the "Water for Life Decade." Presently 2.6 billion people worldwide, 40 percent of the world's population, do not have adequate access to clean water and lack basic sanitation. As a result, thousands of children die every day from diarrhea, cholera, and other hygiene-related diseases. Many young girls miss getting an education because they must spend their time carrying water from distant wells back to the village. The Millennium Developmental Goal of the UN is to cut the proportion of people without sustainable access to clean water and basic sanitation in half by 2015. Building an infrastructure to bring potable water to a billion people probably could be done within ten years, but it would require more resources and resolve from the wealthy nations than have ever been applied to this problem before. Moreover, there just isn't enough water in some parts of the world to provide for the needs of all. Periodic droughts and population growth have led to water scarcity in large regions. According to predictions, by 2025 two-thirds of the world population will be living under water-stress conditions unless something is done. Since most poor people choose food over health, a balance has to be made between sanitation and irrigation that considers the social impact of water allocation policies. In a period of rapid change, how can the many conflicting water interests be balanced equitably?

The United Nations proclaimed a "Sanitation Decade" back in the 1980s, emphasizing the plight of women, but made little headway. The aim was to empower women and bring them into the management of water and sanitation facilities. However, women continued to be relegated to fetching and storing water, and few were appointed to water boards. In many parts of Asia and Africa, the primary schools have no toilets for boys or girls and no hand-washing facilities. Older girls often drop out of school for lack of privacy. In the parts of the new megacities that lack sewers, refuse runs in the streets. However, connecting the urban masses to sewer systems that empty into rivers could be disastrous for those downstream. Moreover, flush sewers use a lot of water, are expensive to install, and deprive farm soils of the nutrients from sewage. An alternate plan, called Ecological Sanitation, or Ecosan (nice euphemism), involves composting waste from shanty towns and using it to fertilize fields after it is free of pathogenic organisms. Perhaps composting should be considered an agricultural activity rather than a waste disposal process. It would recycle needed nutrients and avoid spreading disease. The Japan Toilet Association produces systems that collect urine, feces, and wash water and then compost them into fertilizer. Some even ferment the sewage to produce methane for heating, but such toilets are prohibitively expensive.

Megacities also produce huge amounts of nonbiodegradable solids—old clothes, plastic bottles, tin cans, broken glass, old cars and tires, steel drums containing chemical wastes, electrical cables, and other rubbish. As megacities get larger, the refuse will continue to pile up at ever increasing rates. Garbage dumps will become blighted urban mountains leaking polluted effluents and breeding flies and rats. At some point it will become intolerable. We are in danger of drowning in our own refuse.

POPULATION CONTROL

We seem to have gotten ourselves into quite a pickle. There is barely enough food and water for the 6.6 billion people on earth, and the refuse is piling up. We are contributing to global warming that will lead to rising sea levels and shifts in the growing seasons. Fossil fuels and metal ores are running out (Gordon, Bertram, and Graedel 2006). Wilderness regions are mostly gone or fenced into parks. And the population keeps increasing.

The only solution to all these problems that I can see is to return to the population level of a hundred years ago, when things were more in balance. It would solve overcrowding in cities, deforestation of the hills, pollution of the valleys, overfishing, and overgrazing, and it would reduce the amount of carbon dioxide spewed into the atmosphere. International tensions over dwindling resources would subside, and we might come to realize that we all have to cooperate to live together on this planet. We could return the earth to our descendants in better shape than we found it.

Somehow people all over the world have to become convinced that the planet cannot support all of the more than 6 billion people on it, and that we must quickly reduce the birthrate to avoid disaster. For many years I have been convinced that we are far beyond the carrying capacity of the planet and in danger of a massive die-off, either from hunger, disease, or warfare. As trust and cooperativity dwindle, a sudden breakdown in commerce could lead to famines affecting hundreds of millions of people and might result in nuclear warfare that could turn large areas into radioactive wastelands. Reducing the global population from 6 billion to 2 billion people will be painful and difficult by any means, but some ways are better than others. One of the worst ways is to wait until limiting resources lead to widespread chaos and anarchy and the sorry death of billions by hunger and disease.

A more humane approach to reducing the population could come about through voluntary limits on childbearing. Total fertility rates in Europe, Japan, and North America have dropped significantly in the last thirty years and are now below the replacement level of 2.1. Unfortunately, present governments see this as a problem rather than a welcome state of affairs that should be further encouraged. They consider it a disaster, in which a shrinking number of citizens in the wage-earning years has to support an increasing proportion of retired people. Many governments actively encourage their citizens to have more babies by providing cash "baby bonuses" and free child care. National economic policies have been based on growth for so long that societies are unwilling and unable to adjust to the new realities. They will have to see the bigger picture and embrace the idea that population reductions are essential. Meanwhile, in Africa, the Near East, and the Indian subcontinent, the average family size is still more than 3 children. In some areas, couples have 6 or more children. Although these are some of the poorest countries in the world, their populations keep increasing and adding to the global population. Their governments have to be convinced that every human born into the world is another mouth to feed, another consumer, another burden on the planet. If their aspirations for their nations are to be even partly fulfilled, global population has to decrease rapidly.

With a worldwide policy of "one child at most," population growth could be curbed within a decade and the planetary burden gradually reduced over several generations. During this time, there would be serious shocks to every society as the proportion of young adults dropped dramatically. There would be fewer mouths to feed, but towns and cities would shrink and some would decay. Many women would be freed from child care and would enter the workforce. However, they would expect a voice in societies that have long ignored them. Old people might have to work longer and lower their expectations. I have been told that almost everyone will reject this solution and just let nature take its course. But I am convinced that if more people knew the facts and considered the alternatives, they would see there is no other choice. Many of the values and goals of the last few thousand years would have to be jettisoned to respond to this crisis, but they are the products of another time. Nonconforming countries would have to be brought into line by persuasion, rewards, or sanctions, because this is a problem for all humankind.

India is the largest democracy in the world, with a population of 1.1 billion. As early as 1960, when the population was half what it is now, the government realized that any economic gains it made would be cancelled by population growth, and that drastic measures had to be taken to reduce the birthrate. Rather than educating women and providing birth control devices such as condoms, the government agencies encouraged surgical steriliza-tion. Villagers were given television sets if they convinced enough of their neighbors to have their tubes tied. A state of emergency was declared in 1976, and forced sterilization was implemented in poor neighborhoods. Jobs and government loans were offered only to those who could show that they had been sterilized. Not surprisingly, the Indians did not take kindly to these policies and voted Prime Minister Indira Gandhi out of office the following year. The government gave up on population control for the next twenty years, and the population grew 2 percent each year. Today in India, there is less than 0.1 hectare of land per person, and the cities are teaming. By 2050 India is predicted to overtake China as the world's most populous nation. Clearly, draconian population control does not work in a democracy.

However, it has worked fairly well for the top-down government of China. Vigorous propaganda for birth control started in 1956, when the leadership recognized population growth as an obstacle to development. It was interrupted a few years later by the Great Leap Forward and the Cultural Revolution, which hurt everything except the birthrate. In 1972 the State Council organized committees to oversee birth control activities. "Barefoot doctors" distributed contraceptives to the communes and tried to convince the peasants to limit their children to three or four. In 1979 the government advocated a limit of one child per couple for both rural and urban populations, although members of small ethnic groups were allowed two or three. Couples with no more than one child were given certificates entitling them to cash bonuses, maternity leave, child care, and a better choice of housing. In return they had to pledge that they would have no more children. Cadre leaders would visit the homes of their team members and determine what methods of contraception they were using and whether they became pregnant. Young people were persuaded to postpone marriage, and those with unauthorized pregnancies were pressured to have abortions. Couples with more than one child were exhorted to seek sterilization.

The one-child policy worked better in urban areas than in rural areas, because city dwellers received retirement pensions and did not depend so much on their children in old age. The demography of China has changed as a result of the reduction in births; the largest cohort is now in the twenty-five-to-thirty-year bracket, and 15 percent of the population is over sixty. It is said that the one-child policy has led to a generation of spoiled children who received all the attentions of their parents and missed out on sibling rivalry; however, there are no objective analyses on changes in the national character.

These strong, almost coercive, measures have slowed the rate of increase in the Chinese population; it is expected to peak at 1.5 billion in 2030 and then start to gradually decline. This assumes that sociopolitical and environmental conditions stay about the same, and that social unrest as a result of the growing gap between the rich and the poor does not put them way off course.

Many Muslim and African countries take a dim view of population control. They see efforts to spread birth control as a continuation of colonial exploitation and feel they would command more respect and gain greater stature with larger rather than smaller populations. Their chronic food shortages are thought to be temporary problems that will disappear when a technological fix is in place, but there is none in sight. The population of Pakistan now stands at 165 million people and is expected to double in the next fifty years. It is a hot, dry country that is mostly desert. Potable water is already limited, and the problems will only intensify when twice as many drink from the same wells. Limited water is not the problem in West Africa, but ethnic strife, disease, and corruption are endemic. Nigeria is only slightly larger than Pakistan and has almost the same number of people. The population in Nigeria is also expected to double in fifty years, reaching 340 million by 2050. While this oil-rich nation has benefited from recent increases in the price of crude oil, subsistence agriculture has failed to keep up with rapid population growth. Once a large food exporter, it must now import food paid for with petrodollars. What will happen when the oil runs out later this century?

If there is any hope of reducing global population by family planning, the goal of a smaller population must be strongly supported by almost everyone. A radical change in public opinion is needed, especially in the underdeveloped Third World. People will have to realize that we are way beyond the carrying capacity of the planet: we are engaged in deficit spending, and bankruptcy is down the road. First, all knowledgeable, educated leaders will have to be convinced that it is to their own benefit, as well as the benefit of the whole world, to do everything they can to reduce the population of their countries—not just limit growth but actually reduce the number of people by limiting the number of new births. The facts are clear and the projections irrefutable that continuation on our present course is disastrous. When leaders come to realize that population pressure, even at its present level, will destroy civilization much faster and more thoroughly than global warming or rising sea levels, they may join in the common pursuit of the only sensible course.

Birth control technology is available that can do the job, and it could be further improved by focusing studies on male contraception. Financial assistance in reaching an agreed-upon goal would be provided by wealthier nations as the magnitude of the problem became fully appreciated. Money is not the major obstacle. It is the will that is missing. Assuming that the leaders become fully convinced that population reduction is imperative, they still have to put the policies into practice. How many politicians will go to their people with a plan that goes against the grain of centuries of tradition? They have to work with their people to change the aspirations of everyone.

The second step is to fundamentally change public opinion concerning population goals. Before rules and regulations are put in place, the populace has to see the ultimate goal as good. Public opinion is a fairly recent concept that only arose after the widespread availability of newsletters, books, pamphlets, newspapers, radio, and television. For most of recorded time, the masses seldom had any say in public matters and just went on their traditional way while trying to avoid being run over by armies. Now broadcast news and entertainment inform people of current events and, at the same time, subtly control what they perceive. National fervor to defeat an apparent enemy or to work together to send a person to the moon can be whipped up by the tabloids and public pronouncements. When politicians are thwarted by the checks and balances of government, they often go directly to the people for a mandate. They may color the facts to maximize their chances of getting the desired response, but they know that the voice of the people, in rallies or riots, puts out a strong message. Propaganda and advertising techniques are highly developed and effective. While they are often misused, there is nothing stopping them from being applied to the greatest challenge the human race has ever faced—population reduction. It would not necessarily take that long—five or ten years of concerted effort—to radically change the wishes and desires of the public. We need to be convinced that the directive given after the biblical flood to "be fruitful, and multiply, and replenish the earth" has been fulfilled with a vengeance, and that the present directive is to reduce and repair the damage that has been done.

When the will of the leaders is paired with the will of the people, great things can be accomplished. The first steps are relatively easy. Cooperating to reach an agreed-upon goal is its own reward, and people can take pride in not adding more mouths around the diminishing table. A frown at those who acted selfishly and had two or more children would discourage others from having children. They could talk of the day in the future when children would once again run around a less crowded playground, and the life of all creatures could be sustained.

WONDERFUL LIFE

NO MATTER WHAT HAPPENS IN THE NEXT FEW CENTURIES, LIFE WILL go on. The bacteria that make up most of the biomass will hardly be affected. Some may thrive in newly melted polar regions that have long been buried under glaciers, and some may be challenged as their normal hosts disappear, but the great majority will continue to grow no matter what. However, bacteria interest us less than charismatic large species, especially fellow human beings. Pandas may go extinct in the wild, and the African plains may be stripped of elephants, gorillas, and lions, but rats and mice will probably do all right. The human population will surely shrink, and the way of life for all will be more tightly constrained. But *Homo sapiens* is a resilient species and will adapt in ways yet unknown. Generations may suffer knowing that billions of people are starving to death, and that there is nothing they can do. Slowly a new equilibrium will be found where life can proceed in a sustainable manner. I hope that our great-great-grandchildren will be able to hear the thundering hoofs of wildebeest on the Serengeti.

While the future may not look rosy for humankind, it is important to keep in mind all the advantages and freedoms that we enjoy these days. Those lucky enough to live in wealthy countries seldom have to fear hunger, plague, or violence and have unprecedented opportunities for the love and adventure of life. Some may be stuck in dead-end jobs with no satisfying self-expression, but that is usually a matter of choice. Those who want to be artists at any cost can always chose to be starving artists. The play of symmetry or the colorful line, the peace of a well-tended garden, and a loaf of bread might be enough. Others may find meaningful work that actually pays for the groceries. The arts and sciences are flourishing—music, dance, poetry, genetics, and population biology are more exciting every year. New insights on the evolution of life over the last 4 billion years surprise and humble us constantly. We have come to realize that we are just one branch on the bush of life, but we can appreciate our branch all the more by empathizing with all of life.

The new biology has brought new ways to think of human health and happiness. We know ourselves better, not just the order of DNA bases in our chromosomes, but also our development from a blastocyst to an embryo, to a newborn, to an adult. We have a better appreciation of the genes that we inherited from our distant ancestors and the genes that make us unique. We can peer into the brain and see signs of thought, emotion, and memory. Consciousness is no longer just a mysterious feeling but the subject of experimental science. We can even attempt to determine the neurological basis for good and evil as it is played out in our behavior. Throughout this voyage of discovery, it is important not to forget our humanity. In his old age the great magical realist Gabriel García Márquez wrote a farewell letter to his friends. Its ending makes a fitting end to this book:

I have learned that everyone wants to live on the
peak of the mountain,
without knowing that real happiness is in how it
is climbed.
I have learned that when a newborn child squeezes
for the first time
with his tiny fist his father's finger,
he has him trapped forever.
I have learned that a man has the right to look
down on another only
when he has to help the other get to his feet.

REFERENCES

Adolphs R, Tranel D, Koenigs M, Damasio AR. 2005. Preferring one taste over another without recognizing either. *Nat. Neurosci.* 8:860–861.

Allen JS, Bruss J, Damasio H. 2005. The aging brain: The cognitive reserve hypothesis and hominid evolution. *Am. J. Human Biol.* 17:673–689.

Anjard C, Loomis WF. 2006. GABA induces terminal differentiation of *Dictyostelium* through a GABAB type receptor. *Development* 113:2253–2261.

Arensburg B, Tillier AM, Vandermeersch B, Duday H, Schepartz LA, Rak Y. 1989. A Middle Palaeolithic human hyoid bone. *Nature* 338:758–760.

Armakolas A, Klar A. 2006. Cell type regulates selective segregation of mouse chromosome 7 DNA strands in mitosis. *Science* 311:1146–1149.

Blattner F, Plunkett G 3rd, Bloch CA, Perna NT, Burland V, Riley M, Collado-Vides J, Glasner JD, Rode CK, Mayhew GF, Gregor J, Davis NW, Kirkpatrick HA, Goeden MA, Rose DJ, Mau B, Shao Y. 1997. The complete genome sequence of *Escherichia coli* K-12. *Science* 277:1453–1474.

Bowles S, Gintis H. 2003. Origins of human cooperation. In *Genetic and cultural evolution of cooperation*, ed. P. Hammerstein, pp. 429–444. Cambridge, MA: MIT Press.

Briggs R, King TJ. 1952. Transplantation of living nuclei from blastula cells into enucleated frogs' eggs. *Proc. Natl. Acad. Sci.* 38:455–463.

Camerer CF, Fehr E. 2006. When does "economic man" dominate social behavior? *Science* 311:47–52.

Cann RL, Stoneking M, Wilson, AC. 1987. Mitochondrial DNA and human evolution. *Nature* 325:31–36.

Cello J, Paul A, Wimmer E. 2002. Chemical synthesis of poliovirus cDNA: Generation of infectious virus in the absence of natural template. *Science* 297:1016–1018.

Chou H, Takematsu H, Diaz S, Iber J, Nickerson E, Wright KL, Muchmore EA, Nelson DL, Warren ST, Varki A. 1998. A mutation in human CMP-sialic acid hydroxylase occurred after the *Homo-Pan* divergence. *Proc. Natl. Acad. Sci.* 95:11751–11756.

Cohen J. 1994. *How many people can the Earth support?* New York: Norton and Company.

Daeschler EB, Shubin NH, Jenkins FA. 2006. A Devonian tetrapod-like fish and the evolution of the tetrapod body plan. *Nature* 440:757–763.

Damasio A. 1999. *The feeling of what happens.* New York: Harcourt Brace.

de Quervain D, Fischbacher U, Treyer V, Schellhamme RM, Schnyder U, Buck A, Fehr E. 2004. The neural basis of altruistic punishment. *Science* 305:1254–1258.

Eichinger L, Pachebat JA, Glockner G, Rajandream MA, Sucgang R, et al. 2005. The genome of the social amoeba *Dictyostelium discoideum*. *Nature* 435:43–57.

Fessler D, Haley K. 2003. The strategy of affect: Emotions in human cooperation. In *Genetic and cultural evolution of cooperation*, ed. P. Hammerstein, pp. 7–36. Cambridge, MA: MIT Press.

Foster KR, Shaulsky G, Strassmann JE, Queller DC, Thompson CRL. 2004. Pleiotropy as a mechanism to stabilize cooperation. *Nature* 431: 693–696.

Fraser C, Gocayne JD, White O, Adams MD, Clayton RA, Fleischmann RD, Bult CJ, Kerlavage AR, Sutton G, Kelley JM, Fritchman RD, Weidman JF, Small KV, Sandusky M, Fuhrmann J, Nguyen D, Utterback TR, Saudek DM, Phillips CA, Merrick JM, Tomb JF, Dougherty BA, Bott KF, Hu PC, Lucier TS, Peterson SN, Smith HO, Hutchison CA 3rd, Venter JC. 1995. The minimal gene complement of *Mycoplasma genitalium*. *Science* 270:397–403.

Gagneux P, Cheriyan M, Hurtado-Ziola N, van der Linden EC, Anderson D, McClure H, Varki A, Varki NM. 2003. Human-specific regulation of alpha 2–6-linked sialic acids. *J. Biol. Chem.* 278:48245–48250.

Gazzaniga M. 2005. *The ethical brain.* New York: Dana Press.

Gilbert S. 2006. *Developmental biology.* 8th ed. Sunderland, MA: Sinauer Associates.

Gilbert SF, Tyler A, Zackin E. 2005. *Bioethics and the new embryology.* Sunderland, MA: Sinauer Associates.

Gilbert SL, Dobyns WB, Lahn BT. 2005. Genetic links between brain development and brain evolution. *Nat. Rev. Genet.* 6:581–590.

Gordon R, Bertram M, Graedel T. 2006. Metal stocks and sustainability. *Proc. Natl. Acad. Sci.* 103:1209–1214.

Hammerstein P. 2003. Why is reciprocity so rare in social animals? In *Genetic and cultural evolution of cooperation*, ed. P. Hammerstein, pp. 83–94. Cambridge, MA: MIT Press.

Hardin, G. 1968. The tragedy of the commons. *Science* 162:1243–1248.

Hsu M, Bhatt M, Adolphs R, Tranel D, Camerer CF. 2005. Neural systems responding to degrees of uncertainty in human decision-making. *Science* 310:1680–1683.

Huber C, Eisenreich W, Hecht S, Wachtershauser G. 2003. A possible primordial peptide cycle. *Science* 301:938–940.

Kimura K, Ote M, Tazawa T, Yamamoto D. 2005. Fruitless specifies sexually dimorphic neural circuitry in the *Drosophila* brain. *Nature* 438:229–233.

Kishigami S, Mizutani E, Ohta H, Hikichi T, Thuan N, Wakayama S, Bui H, Wakayama T. 2006. Significant improvement of mouse cloning technique by treatment with trichostatin A after somatic nuclear transfer. *Biochem. Biophys. Res. Commun.* 340:183–189.

Kitcher P. 1984. *Vaulting ambition: Sociobiology and the quest for human nature.* Cambridge, MA: MIT Press.

———. 1996. *The lives to come.* New York: Simon and Schuster.

———. 2001. *Science, truth, and democracy.* Oxford: Oxford Univ. Press.

———. 2007. *Living with Darwin: Evolution, design, and the future of faith.* New York: Oxford Univ. Press.

Klar AJ. 2004. An epigenetic hypothesis for human brain laterality, handedness, and psychosis development. *Cold Spring Harbour Symp. Quant. Biol.* 69:499–506.

———. 2005. A 1927 study supports a current genetic model for inheritance of human scalp hair-whorl orientation and hand-use preference traits. *Genetics* 170:2027–2030.

Knoll A. 2003. *Life on a young planet.* Princeton: Princeton Univ. Press.

Koch C. 2003. *The quest for consciousness: A neurobiological approach.* Engelwood, CO: Roberts and Company.

Kopp RE, Kirschvink JL, Hilburn IA, Nash CZ. 2005. The Paleoproterozoic snowball Earth: A climate disaster triggered by the evolution of oxygenic photosynthesis. *Proc. Natl. Acad. Sci.* 102:11131–11136.

Lamason R, Mohideen MA, Mest JR, Wong AC, Norton HL, Aros MC, Jurynec MJ, Mao X, Humphreville VR, Humbert JE, Sinha S, Moore JL, Jagadeeswaran P, Zhao W, Ning G, Makalowska I, McKeigue PM, O'Donnell D, Kittles R, Parra EJ, Mangini NJ, Grunwald DJ, Shriver MD, Canfield VA, Cheng KC. 2005. SLC24A5, a putative cation exchanger, affects pigmentation in zebrafish and humans. *Science* 310: 1782–1786.

Leman L, Orgel L, Ghadiri MR. 2004. Carbonyl sulfide–mediated prebiotic formation of peptides. *Science* 306:283–286.

Lewontin R. 1980. Sociobiology: Another biological determinism. *Int. J. Health Serv.* 10:347–363.

Loomis WF. 1975. *Dictyostelium discoideum: A developmental system.* New York: Academic Press.

———. 1986. *Developmental biology.* New York: Macmillan.

———. 1988. *Four billion years.* Sunderland, MA: Sinauer Associates.

Luria S, Delbruck M. 1943. Mutations of bacteria from virus sensitivity to virus resistance. *Genetics* 28:491–511.

Margulis L, Sagan D. 1995. *What is life?* Berkeley: Univ. of California Press.

Matsuoka Y, Furuyashiki T, Yamada K, Nagai T, Bito H, Tanaka Y, Kitaoka S, Ushikubi F, Nabeshima T, Narumiya S. 2005. Prostaglandin E receptor EP1 controls impulsive behavior under stress. *Proc. Natl. Acad. Sci.* 102:16066–16071.

Mekel-Bobrov N, Gilbert SL, Evans PD, Vallender EJ, Anderson JR, Hudson RR, Tishkoff SA, Lahn BT. 2005. Ongoing adaptive evolution of ASPM, a brain size determinant in *Homo sapiens. Science* 309:1720–1722.

Miller SL, Urey H. 1953. Organic compound synthesis on the primitive earth. *Science* 130:245–251.

Moll J, Zahn R, de Oliveira-Souza R, Krueger F, Grafman J. 2005. The neural basis of human moral cognition. *Nat. Rev. Neurosci.* 6:799–809.

Morwood M, Brown P, Jatmiko, Sutikna T, Saptomo EW, Westaway KE, Due RA, Roberts RG, Maeda T, Wasisto S, Djubiantono T. 2005. Further evidence for small-bodied hominins from the Late Pleistocene of Flores, Indonesia. *Nature* 437:1012–1017.

Olcott AN, Sessions AL, Corsetti FA, Kaufman AJ, de Oliviera TF. 2005. Biomarker evidence for photosynthesis during neoproterozoic glaciation. *Science* 310:471–474.

Olsen RM, Loomis WF. 2005. A collection of amino acid replacement matrices derived from clusters of orthologs. *J. Mol. Evol.* 61:659–665.

Paine J, Shipton CA, Chaggar S, Howells RM, Kennedy MJ, Vernon G, Wright SY, Hinchliffe E, Adams JL, Silverstone AL, Drake R. 2005. Improving the nutritional value of Golden Rice through increased provitamin A content. *Nat. Biotechnology* 23:482–487.

Pinker, S. 1997. Evolutionary biology and the evolution of language. In *The origin and evolution of intelligence*, ed. A. Schrieber and J. W. Schopf. Boston: Jones and Bartlett.

Posfai G, Plunkett GR, Feher T, Frisch D, Keil G, Umenhoffer K, Kolisnychenko V, Stahl B, Sharma S, de Arruda M, Burland V, Harcum S, Blattner F. 2006. Emergent properties of reduced-genome *Escherichia coli*. *Science* 312:1044–1046.

Richardson PJ, Boyd RT, Henrich J. 2003. Cultural evolution of human cooperation. In *Genetic and cultural evolution of cooperation*, ed. P. Hammerstein, pp. 357–388. Cambridge, MA: MIT Press.

Rilling J, Gutman D, Zeh T, Pagnoni G, Berns G, Kilts C. 2002. A neural basis for social cooperation. *Neuron* 35:395–405.

Ronshaugen M, McGinnis N, McGinnis W. 2002. Hox protein mutation and macroevolution of the insect body plan. *Nature* 415:914–917.

Rose S. 1992. *The making of memory*. New York: Doubleday.

Schnieke A, Kind AJ, Ritchie WA, Mycock K, Scott AR, Ritchie M, Wilmut I, Colman A, Campbell KH. 1997. Human factor IX transgenic sheep produced by transfer of nuclei from transfected fetal fibroblasts. *Science* 278:2130–2133.

Schroedinger E. 1944. *What is life?* Cambridge: Cambridge Univ. Press.

Senanayake S, Idriss H. 2006. Photocatalysis and the origin of life: Synthesis of nucleoside bases from formamide on $TiO_2(001)$ single surfaces. *Proc. Natl. Acad. Sci.* 103:1194–1198.

Seyfarth RM, Cheney DL. 1997. Communication and the minds of monkeys. In *The origin and evolution of intelligence*, ed. A. Schrieber and J. W. Schopf. Boston: Jones and Bartlett.

Shea JB. 2006. Catholic teaching on the human embryo as an object of research. Catholic Insight. December 3, http://catholicinsight.com/online/bioethics/embryo.shtml, accessed on June 9, 2007.

Shermer M. 2004. *The science of good and evil.* New York: Holt and Company.

Shumyatsky G, Malleret G, Shin R, Takizawa S, Tully K, Tsvetkov E, Zakharenko S, Joseph J, Vronskaya S, Yin D, Schubart U, Kandel E, Bolshakov V. 2005. Stathmin, a gene enriched in the amygdala, controls both learned and innate fear. *Cell* 18:697–709.

Song J, Olsen R, Loomis WF, Shaulsky G, Kuspa A, Sucgang R. 2005. Comparing the *Dictyostelium* and *Entamoeba* genomes reveals an ancient split in the *Conosa* lineage. *PLoS Comput. Biol.* 1:579–584.

Strassmann JE, Zhu Y, Queller DC. 2000. Altruism and social cheating in the social amoeba *Dictyostelium discoideum. Nature* 408:965–967.

Taylor AL, Trotter CD. 1967. Revised linkage map of *Escherichia coli. Bacteriol. Rev.* 31:332–353.

Thomson J, Itskovitz-Eldor J, Shapiro S, Waknitz M, Swiergiel J, Marshall V, Jones J. 1998. Embryonic stem cell lines derived from human blastocysts. *Science* 303:1674–1677.

Varki A, Altheide TK. 2005. Comparing the human and chimpanzee genomes: Searching for needles in a haystack. *Genome Res.* 15:1746–1758.

Wakayama T, Perry AC, Zuccotti M, Johnson KR, Yanagimachi R. 1998. Full-term development of mice from enucleated oocytes injected with cumulus cell nuclei. *Nature* 394:369–374.

Wakayama T, Tabar V, Rodriguez I, Perry AC, Studer L, Mombaerts P. 2001. Differentiation of embryonic stem cell lines generated from adult somatic cells by nuclear transfer. *Science* 292:740–743.

Weber B, Hoppe C, Faber J, Axmache N, Fliessbach K, Mormann F, Weis S, Ruhlmann J, Elger CE, Fernandez G. 2006. Association between scalp hair-whorl direction and hemispheric language dominance. *Neuroimage* 30:539–543.

Wehner R. 1997. Prerational intelligence—how insects and birds find their way. In *The origin and evolution of intelligence*, ed. A. Schrieber and J. W. Schopf. Boston: Jones and Bartlett.

White T, Wolde Gabriel G, Asfaw B, Ambrose S, Beyene Y, Bernor RL, Boisserie JR, Currie B, Gilbert H, Haile-Selassie Y, Hart WK, Hlusko LJ, Howell FC, Kono RT, Lehmann T, Louchart A, Lovejoy CO, Renne PR, Saegusa H, Vrba ES, Wesselman H, Suwa G. 2006. Asa Issie, Aramis, and the origin of *Australopithecus*. *Nature* 440:883–889.

Wilmut I, Schnieke A, McWhir J, Kind A, Campbell K. 1997. Viable offspring derived from fetal and adult mammalian cells. *Nature* 385:810–813.

Wilson EO. 1975. *Sociobiology: The new synthesis.* Cambridge, MA: Harvard Univ. Press.

Wittlinger M, Wehner R, Wolf H. 2006. The ant odometer: Stepping on stilts and stumps. *Science* 312:1965–1967.

Xiao D, Houser D. 2005. Emotion expression in human punishment behavior. *Proc. Natl. Acad. Sci.* 102:7398–7401.

Zhang Y, Lu H, Bargmann CI. 2005. Pathogenic bacteria induce aversive olfactory learning in *Caenorhabditis elegans. Nature* 438:179–184.

Zhao S, Maxwell S, Jimenez-Beristain A, Vives J, Kuehner E, Zhao J, O'Brien C, de Felipe C, Semina E, Li M. 2004. Generation of embryonic stem cells and transgenic mice expressing green fluorescence protein in midbrain dopaminergic neurons. *Eur. J. Neurosci.* 19:1133–1140.

INDEX

Numbers in italics refer to figures.

abortion, 78–79, 87–92
adenosine deaminase (ADA), 53–54, 55
agricultural revolution, 210–12
AIDS, 41–42, 67
albinism, 69–70
altruism, 152; kin selection and, 161; reciprocal, 153, 155, 165
Alzheimer's disease, 144
amino acids, 5, 6, 10
amphibians, 196
anger, 166
Antinori, Severino, 63–64
Aristotle, 134–35
aspartoacylase (ASPA), 86–87
ATP, 11, 12, 14
Australopithecus: afarensis, 180, 197–98; *africanus,* 198

bacteria: aerobic, 192; anaerobic, 190–91; early, 184, 189; engineered circuits in, 75–76; life is

cheap for, 16; shared common ancestry with humans of, 5–6, 10–11; signs of life in, 11
biodiesel, 219
Blackmun, Harry A., 20
Boltwood, Bertram, 8–9
brain, the: aging of, 144; awareness and, 130–33; language and, 131; mutations and evolution of, 140–41; size genes, 141; structure of, *118,* 129–34; taste and, 133–34
Broca's area, 131
Bt proteins, 58

Caenorhabitis elegans, 109–10, 115–16
Canavan disease, 86–87
carrying capacity, 211, 212, 224, 229
Catholic Church, 24, 31, 88–89
cells, 4–5; are cheap, xii; gene expression in, 28–29. *See also* embryonic stem (ES) cells; somatic stem cells

Text: 10/15 Janson
Display: Interstate
Compositor: BookComp, Inc.
Indexer: Andrew Christenson